Science and Technology Policy Institute

E-Vision 2000:
Key Issues That Will Shape Our Energy Future

Summary of Proceedings, Scenario Analysis, Expert Elicitation, and Submitted Papers

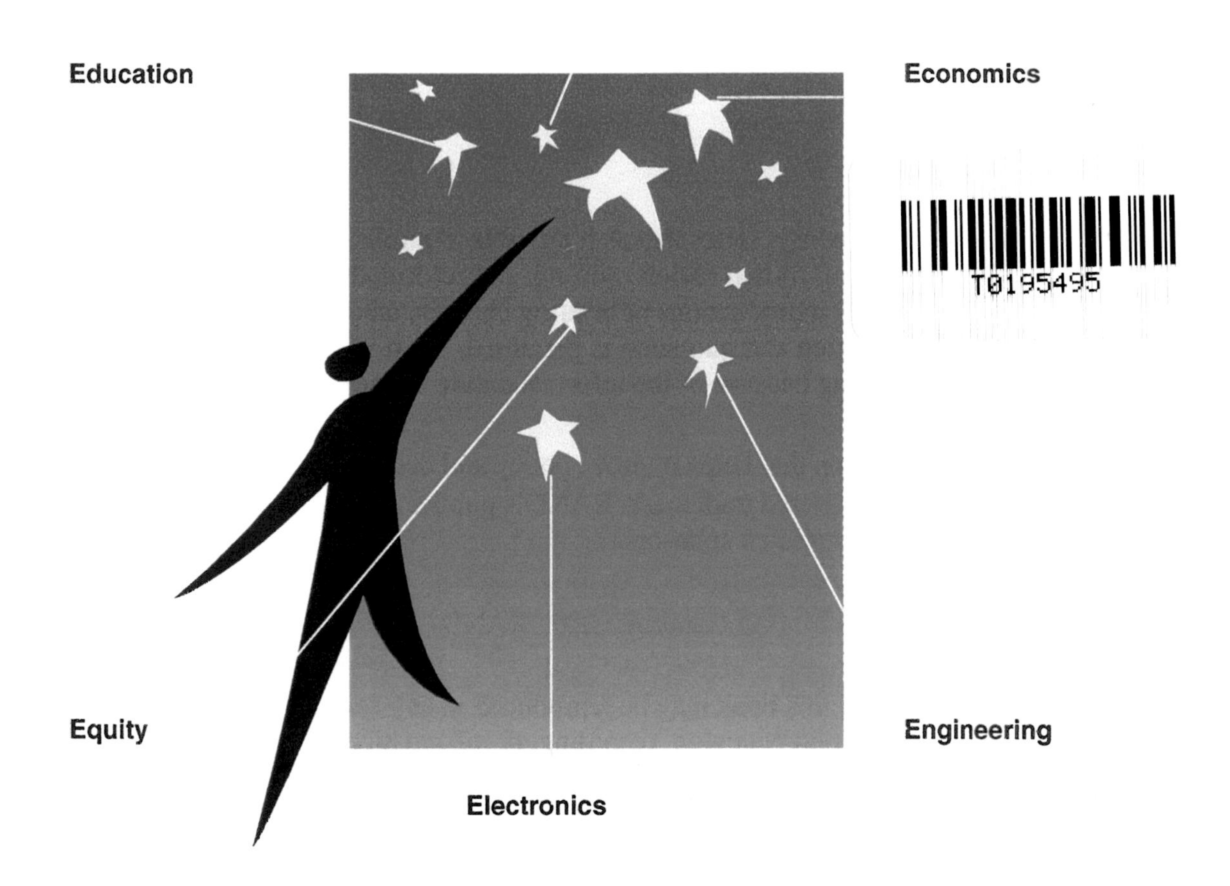

Prepared for the Department of Energy

CF-170-DOE

June 2001

RAND

The research described in this report was conducted by RAND's Science and Technology Policy Institute under Contract ENG-9812731.

ISBN: 0-8330-3054-X

The RAND conference proceedings series makes it possible to publish conference papers and discussions quickly by forgoing formal review, editing, and reformatting. Proceedings may include materials as diverse as reproductions of briefing charts, talking points, or carefully written scientific papers. Citation and quotation is permitted, but it is advisable to check with authors before citing or quoting because of the informal nature of the material.

RAND is a nonprofit institution that helps improve policy and decisionmaking through research and analysis. RAND® is a registered trademark. RAND's publications do not necessarily reflect the opinions or policies of its research sponsors.

Published 2001 by RAND
1700 Main Street, P.O. Box 2138, Santa Monica, CA 90407-2138
1200 South Hayes Street, Arlington, VA 22202-5050
201 North Craig Street, Suite 102, Pittsburgh, PA 15213
RAND URL: http://www.rand.org/
To order RAND documents or to obtain additional information, contact Distribution Services: Telephone: (310) 451-7002; Fax: (310) 451-6915; Email: order@rand.org

Preface

This report documents a May 2000 initiative by the Office of Energy Efficiency and Renewable Energy (EERE) of the U.S. Department of Energy (DOE) to identify and assess a range of emerging issues that may affect future energy use and supply. EERE contracted with RAND's Science and Technology Policy Institute to plan and execute a strategic planning project. The project's purpose was to explore possible new approaches to energy supply and use, identify key issues EERE will face, and consider the implications for EERE's research and development (R&D) portfolio. EERE was interested in developing a systematic process to better understand the implications of emerging issues for future R&D, federal assistance programs, public information dissemination, education, and technical training.

The project had three parts: (1) a conference called E-Vision 2000, held October 11–13, 2000 in Washington, D.C., including presentation of invited papers; (2) an assessment of long-range planning scenarios currently used in the energy community; and (3) a structured process to identify a set of critical energy issues in 2020 to inform the EERE R&D portfolio, as viewed by a range of energy experts.

This report summarizes the issues raised and suggestions made for future research by the participants in and attendees of the E-Vision 2000 conference. It also summarizes the key insights derived from RAND's scenario analysis and expert elicitation and includes abstracts of papers some of the panelists submitted.

RAND played the roles of conference convener, organizer, and integrator and compiled summaries of issues and comments. However, it should be noted that this report is not intended to reflect RAND's thinking on the subject of energy R&D but rather the views of the many participants in the E-Vision 2000 process.

The S&T Policy Institute

Originally created by Congress in 1991 as the Critical Technologies Institute and renamed in 1998, the Science and Technology Policy Institute is a federally funded research and development center sponsored by the National Science Foundation and managed by RAND. The Institute's mission is to help improve public policy by conducting objective, independent research and analysis on policy issues that involve science and technology. To this end, the Institute:

- Supports the Office of Science and Technology Policy and other Executive Branch agencies, offices, and councils;
- Helps science and technology decisionmakers understand the likely consequences of their decisions and choose among alternative policies; and
- Helps improve understanding in both the public and private sectors of the ways in which science and technology can better serve national objectives.

Science and Technology Policy Institute research focuses on problems of science and technology policy that involve multiple agencies. In carrying out its mission, the Institute consults broadly with representatives from private industry, institutions of higher education, and other nonprofit institutions.

Inquiries regarding the Science and Technology Policy Institute may be directed to the addresses below.

Bruce Don
Director

Science and Technology Policy Institute
RAND
1200 South Hayes Street
Arlington, Virginia 22202-5050
Phone: (703) 413-1100 x5351
Web: http://www.rand.org/scitech/stpi
Email: stpi@rand.org

Contents

Summary

EERE contracted with RAND's Science and Technology Policy Institute to plan and execute a strategic planning exercise. The project's purpose was to explore possible new approaches to energy supply and use, identify key issues EERE will face, and consider the implications for EERE's R&D portfolio. The project had three parts: (1) the E-Vision 2000 Policy Forum, held October 11–13, 2000 in Washington, D.C.; (2) an assessment of planning scenarios currently used in the energy community; and (3) a structured process to identify critical energy issues in 2020 to inform the EERE R&D portfolio, as viewed by a range of experts.

A key organizing principle of the E-Vision 2000 process was the concept of energy productivity. Just as the productivity of labor and capital is used to measure how well those markets function, energy productivity can be used as an indicator of the health of energy markets. Energy productivity is the ratio of gross domestic product to total energy consumption in the economy. It captures the sum total of energy used to produce, store, distribute, use, and dispose of goods and services. Inefficiency in overall energy use in the economy would be reflected as low energy productivity. In the last decade, energy productivity has taken on even greater significance in the economy as information technologies (IT) have begun to transform business and consumer patterns of energy use. Specific examples of this ongoing transformation were discussed at the E-Vision 2000 Policy Forum and explored in the analysis of alternative energy scenarios.

E-Vision 2000 Policy Forum

The E-Vision 2000 Policy Forum was built around three expert panels and a final open session. The three main sessions included presented papers and panel discussions on the following issues:

- Role of information technologies in shaping future energy needs and use
- Factors shaping worker and student productivity in buildings and the relationship to energy use
- Effects of applying a "systems" approach to current and future energy supply and demand practices.

Each panel was chaired by a distinguished member of the energy community and included several energy experts who prepared and delivered papers relevant to the panel topics. Several keynote speakers also presented their views on important issues. On the third day of the conference, an open forum was held to provide conference attendees with an opportunity to bring up their own issues. The speakers and other forum participants identified research opportunities in areas that have either not been initiated or that need further development.

Influence of Information Technologies on Energy Use

This session was chaired by Henry Lee, Director, Environment and Natural Resources Program, Harvard University. The panel addressed the several questions, such as the following:

- How might new information technologies (IT) be used to manage energy demand and use?
- What effects will e-commerce and telecommuting have on energy demand?
- How will information technologies affect energy consumption?
- Is net consumption likely to increase or decrease?
- What federal research could help develop methods to capture the potential energy savings of information technologies?

Several general points were made regarding a future research agenda:

- Little is currently known about the relationship between IT and energy systems.
- IT may be changing the relationship between energy use and economic growth, but the dimensions of change in energy productivity are still not well understood.
- Improved electricity reliability and quality are critical issues for an IT-based economy.
- While IT makes real-time information on energy use and prices possible, little is known about consumer and business response to this information, the necessity of government standards for accuracy, and the costs and benefits of providing this information.
- IT is having significant effects on development of competitive energy markets.

Worker Productivity, Building Design, and Energy Use

This panel was chaired by Vivian Loftness, Chair of the Department of Architecture, Carnegie Mellon University. The panel examined the following questions:

- What opportunities exist for improvements in building design to enhance student and worker productivity and health and to decrease energy use?
- To what extent is productivity of students and workers enhanced through improvements in building energy efficiency and indoor air quality?

Several key points were raised regarding future research needs:

- An interdisciplinary research program is required for the study of worker health and productivity, the role of building design in worker health and productivity, and the implications for energy use.
- Effectively addressing innovation in building design requires a systems approach, possibly spanning multiple federal agencies. However, it is not clear who should lead this research and whether researchers have the capabilities to fully pursue a systems approach.[1]

Panelists suggested that DOE could play a decisive role in initiating and coordinating a research agenda in these areas. For example, DOE could

- support the study of the relationships between worker productivity and building environments and their impacts on energy use
- act as a clearinghouse of information on the research findings from various disciplines to encourage spillovers in knowledge to other disciplines
- provide demonstration sites of buildings or parts of buildings to test research results and to educate researchers, architects, and building developers on the economics of incorporating building design changes.

Systems Approach to Energy Use

Chaired by Maxine Savitz, General Manager for Technology Partnerships at Honeywell International, this panel addressed several key questions, such as the following:

[1]The term *systems approach* refers here to a multidisciplinary study of building design that explicitly deals with connections to other economic activities, such as worker productivity and energy use.

- What technical advances are needed to realize potential energy savings or productivity enhancements from systems-based and performance-based approaches to managing energy use?
- What practical, computer-based tools are needed for a cradle-to-grave approach to system optimization from conceptualization through operation and retirement?
- What metrics should be used to measure energy efficiency and productivity and decreases in energy-related pollution, especially in competitive electricity markets?

In this context, the term *systems approach* refers to an integrated view of energy production, distribution, and use and their explicit connections to other economic and social activities. The panel noted that the institutions, markets, and technologies that define the energy industry are undergoing massive change, domestically and globally. A systems approach would substantially improve the quality of the U.S. response to these challenges and opportunities. For example,

- The nation has many policy, regulatory, and financial tools to influence energy systems, but does not have a means of coordinating the use of these tools very well.
- National rules regarding interconnection standards, emergency routing procedures, and other matters could facilitate the movement toward restructuring the electricity industry.
- Firms are already installing distributed generation, but regulatory and technical obstacles cause these systems to be isolated from the grid.
- Implementing distributed generation will require new technologies and techniques to ensure reliability and optimize operation, but these potential changes need to be taken in the context of the energy system as a whole.
- Fossil fuels could be used more efficiently throughout the energy chain including extraction, conversion, and end-use, even as non-fossil technologies continue to be developed.
- Energy conservation, energy efficiency, and renewable energy are integral components of a more sustainable "zero net energy" approach to design.

Open Forum

Conference participants who had not served as panelists or speakers in the Forum were given an opportunity to make brief presentations. Several speakers encouraged DOE to pursue further research on biomass as a fuel source. Other speakers recommended additional research on carbon sequestration in the oceans and hydrogen-based energy systems. In addition, some of the speakers in the Open Forum suggested that DOE needed to take a more active role in stimulating innovative thinking about energy systems and working more closely with educators to expand their vision of a more sustainable energy future.

Scenario Analysis

In the planning phase of the E-Vision 2000 process, the scenario analysis was envisioned to provide conference participants with a common base of understanding the key features of widely available energy scenarios. In practice, the scenario analysis was not integrated into the panel discussions of the Policy Forum because the agenda lacked sufficient time. However, the analysis significantly improved understanding of the essential similarities and differences among the many energy scenarios that have been published in recent years and will be used in follow-on analysis with other conference outputs.

RAND's scenario analysis team defined a set of parameters that could provide a common framework to compare and contrast the large number of commonly used energy planning scenarios in ways that are meaningful for policy development. This framework was then used to identify groupings of individual scenarios that comprised "meta-scenarios," representing a plausible set of alternative futures. These meta-scenarios were assessed to determine signposts, hedging strategies, and shaping strategies needed for an adaptive approach to strategic energy planning. Signposts are critical indicators of changing circumstances. Shaping strategies are actions that can alter future events, while hedging strategies are actions to help cope with events that cannot be altered.

The analysis of these studies resulted in four key observations that illustrate how such scenarios and studies can provide policy insights despite the uncertainty associated with long-term projections:

- Research and development serves as a hedging strategy, providing a means of dealing with technological uncertainties. One policy challenge is to identify R&D portfolios that are robust under a range of plausible scenarios.
- If "decarbonization" policies (i.e., those that result in less use of fossil fuels) are pursued, many of the scenarios examined suggest natural gas as a transition fuel, whether the gas is used by fuel cells or advanced combined-cycle turbines. The policy implication is whether sufficient natural gas supplies could possibly be developed and pipelines and storage

infrastructure constructed in the time frame required by some of the scenarios.

- The Internet and information technology may be changing the way we use energy in fundamental ways. The policy challenge is to understand the mechanism of change to improve projections of demand, use, and productivity.
- Such technologies as fuel cells and distributed generation may lead to fundamental changes in energy systems and productivity through their wide range of potential applications. The key question is how to devise policies that might foster this process of fundamental change.

Additional work is needed to take these scenarios to the point where the implications for energy R&D can be more clearly determined and the portfolio appropriately focused.

Expert Elicitation of Key Issues

Twenty seven high-level national energy thinkers participated in a process to identify the most important issues that will affect energy trends in the next 20 years. The process used was a modified version of the Delphi method, developed at RAND from studies on decisionmaking that began more than 50 years ago. The Delphi method is designed to take advantage of the talent, experience, and knowledge of a group of experts in a structure that allows for an exchange of divergent views. In this application, EERE was seeking a formal means of capturing expert opinion of energy needs in 2020, which would inform R&D portfolio choices over the next five to ten years.

The participants in this Delphi process showed a convergence of interest in the following energy issues:

- Global warming/greenhouse gas emissions (70 percent of the participants)
- Hybrid/zero-emission vehicle market penetration (59 percent of the participants)
- Natural gas usability/use (56 percent of the participants)
- Increased nuclear use (33 percent of the participants
- Oil prices (26 percent of the participants)
- Fuel cell viability (19 percent of the participants).

The results of this Delphi exercise will contribute to planning future research as well as subsequent E-Vision exercises. In follow-on work, views of the future will be sought from non–energy experts knowledgeable about a wide range of economic and social issues. Their insights will then be examined in terms of their implications for energy development and use, and redeployment of R&D priorities for EERE.

Submitted Papers

Prior to the E-Vision 2000 conference, RAND asked speakers to prepare written papers on topics related to their panels' focus area. These papers were sent to participants in advance of the conference, with the intent of providing participants with a common base of information, and adding more substance and focus to the conference proceedings.

Beginning on page 57, this volume provides titles, authors, and abstracts for all the submitted papers. The papers themselves are available in the companion volume, as well as online at **http://www.rand.org/scitech/stpi/Evision/Supplement/** in their entirety.

Note that these papers reflect the views of their authors, not necessarily those of the Department of Energy, RAND, or any other organization with which the authors may be affiliated.

Acknowledgments

Special thanks are owed to the following RAND staff who contributed to the writing, editing, review, and compilation of this material: James Dewar, Bruce Don, Jeff Drezner, John Friel, Phyllis Gilmore, Anders Hove, Scott Hassell, James Kadtke, Debra Knopman, Susan Resetar, Lisa Sheldone, Rich Silberglitt, Sally Sleeper, Katie Smythe, and Jerry Sollinger.

Introduction

In the spring of 2000, the Office of Energy Efficiency and Renewable Energy (EERE) of the Department of Energy (DOE) contracted with RAND's Science and Technology Policy Institute to plan and execute a strategic planning project. The project's purpose was to explore possible new approaches to energy supply and use, identify key issues EERE will face, and consider the implications for EERE's research and development (R&D) portfolio. The project had three parts: (1) the E-Vision 2000 conference, held October 11–13, 2000 in Washington, D.C.; (2) an assessment of planning scenarios currently used in the energy community; and (3) a structured process to identify critical energy issues in 2020 as viewed by a range of experts. In addition, E-Vision 2000 panelists submitted papers in advance of the conference.

Exploring the Dimensions of Energy Productivity

A key organizing principle of the E-Vision 2000 process was the concept of energy productivity. Just as the productivity of labor and capital is used to measure how well those markets function, energy productivity can be used as an indicator of the health of energy markets. Energy productivity is the ratio of gross domestic product to total energy consumption in the economy. It captures the sum total of energy used to produce, store, distribute, use, and dispose of goods and services. Inefficiency in overall energy use in the economy would be reflected as low energy productivity. Its scope is far broader than energy conservation (simple demand reduction) or energy efficiency (using less energy per unit product or service).

From a strategic planning perspective, improving energy productivity is the critical objective that links demand, supply, and economic activity. It includes more than the simple efficiency of electrical devices. For example, more-sophisticated production and use choices are increasingly available through information technology (IT), such as avoiding the production of excess inventory and using automated timers to control heating and air conditioning in buildings.

In the last decade, energy productivity has taken on even greater significance in the economy as IT have begun to transform business and consumer patterns of energy use. Specific examples of this ongoing transformation may be found in this report and the supporting material. E-Vision 2000 participants also explored the implications of changing building designs to improve both worker and energy productivity simultaneously. The panel on systems approaches to energy use discussed examples of improving energy productivity through fundamental changes in transportation systems, land-use planning, and electricity grid design. The panel also noted that only a systems approach could fully measure the impact on energy productivity.

Organization of the Report

This report is intended to be an archival document that reflects the scope and limits of the E-Vision 2000 process. As such, it begins by summarizing the key issues raised within the panels and open forum of the conference. The summaries of issues presented at the E-Vision conference 2000 Policy Forum are derived from several sources:

- panel discussion and questions from the floor (papers, video, and transcripts)
- introductory and lunch speakers (video and transcripts)
- open forum (transcripts).

In the summaries of panel discussions as well as the open forum at the end of the conference, RAND has attempted to communicate the information and viewpoints expressed by the conference participants. The summaries also reflect the free-flowing nature of the conference format. Clarification has been added where necessary to convey the meaning of a statement. Where possible, ideas are attributed to the appropriate participant; see Reference Participants, below, for a full list of individuals.

The report also includes a summary of the energy scenarios and a brief description of the expert elicitation known as the Delphi process. Finally, the report includes abstracts of 13 papers written specifically for the conference, and provided to conference attendees and panel members to stimulate discussion.

See http://www.rand.org/scitech/stpi/Evision/ for transcripts of the E-Vision Conference, the full text of the scenario analysis and submitted papers, and slides outlining the Delphi process.

Referenced Participants

Brad Allenby, Vice President, Environment, Health and Safety, AT&T

Dan E. Arvizu, Group Vice President, Energy and Industrial Systems, CH2MHill

Jerry Asher, Associate Director, Electric Vehicle Association of Greater D.C.

Mark Berman, Principal, Business Development, Davis Energy Group

Jeffrey E. Christian, Director, Buildings Technology Center, Oak Ridge National Laboratory

Jack Cogen, President, Natsource LLC

James Dewar, Senior Mathematician and Professor of Policy Analysis, RAND Graduate School, RAND

Bruce Don, Director, Science and Technology Policy Institute, RAND

Kevin M.D. Dye, Chief Process Scientist, CWA Ltd.

Charles K. Ebinger, Vice President, International Energy Services, Nexant

William Fisk, Staff Scientist, Lawrence Berkeley National Laboratory

Hans Friedericy, Department of Water and Power, City of Los Angeles; General Manager, Honeywell Aerospace; Consultant

Jeffrey P. Harris, Group Leader, Government and Industry Programs, Lawrence Berkeley National Laboratory

Dale Jorgenson, Director, Program on Technology and Economic Policy, John F. Kennedy School of Government, Harvard University

Kevin Kampshroer, Research Director, Public Buildings Service, General Services Administration

Kevin Kephart, Director, Agricultural Experiment Station, South Dakota State University

T. Keith Lawton, Director of Travel Forecasting, Department of Transportation, City of Portland, Oregon

Henry Lee, Director, Environment and Natural Resources Program, Harvard University

Vivian Loftness, Chair, Department of Architecture, Carnegie Mellon University

Michael Markels, President, CEO, Ocean Farming Inc.

Rose McKinney-James, President, Government Affairs, Faiss Foley Merica

William A. McDonough, President, William McDonough & Partners; Professor, Department of Architecture; University of Virginia

Ernest J. Moniz, Under Secretary for Energy, Science and Environment; U.S. Department of Energy

Alvin Pack, Senior Vice President, Energyharbor.com

Stefano Ratti, Assistant Director, Center for Environmental Energy Engineering, University of Maryland

Stephen Rattien, Director, RAND Science and Technology, RAND

William Reed, Vice President and Chief Regulatory Officer, Sempra Energy

Dan W. Reicher, Assistant Secretary of Energy, Energy Efficiency and Renewable Energy, U.S. Department of Energy

William B. Richardson, Secretary of Energy, U.S. Department of Energy

Joseph Romm, Executive Director, Center for Energy and Climate Solutions, Global Environment and Technology Foundation

Arthur H. Rosenfeld, Commissioner, California Energy Commission

Megan Smith, Codirector, Biofuels Transportation Issues, American Bioenergy Association; MSS Consultants, LLC

Robert C. Sonderegger, Director of Modeling and Simulation, Silicon Energy Corporation

Karl Stahlkopf, Vice President, Power Delivery, Electric Power Research Institute

Kevin J. Stiroh, Senior Economist, Federal Reserve Bank of New York

Bruce Stram, Vice President, ENRON Corporation

Darian William Unger, Researcher, Energy Laboratory, Massachusetts Institute of Technology

Matthew Varilek, Analyst, Advisory Services Unit, Natsource LLC

James Winebrake, Associate Professor, James Madison University

Richard Wright, Retired Director, Natural Research Council

David P. Wyon, Research Fellow; Adjunct Professor, Technical University of Denmark

Kurt Yeager, President, Electric Power Research Institute

Influence of Information Technologies on Energy Use

Issue

Will the computer, Internet, and electronics revolution decrease or increase energy Use?

Panel Chair

Henry Lee
Director, Environment and Natural Resources Program, Harvard University

Panelists

Dale Jorgenson
Director, Program on Technology and Economic Policy, Kennedy School of Government, Harvard University

Joseph Romm
Executive Director, Center for Energy and Climate Solutions, Global Environment & Technology Foundation

Brad Allenby
Vice President, Environmental Health and Safety, AT&T

Karl Stahlkopf
Vice President of Power Delivery, Electric Power Research Institute

William Reed
Vice President and Chief Regulatory Officer, Sempra Energy

This panel explored the relationship between various information technologies and energy use. In the following summary, points raised by panelists and others are categorized as issues pertaining to either general research, demand, supply, or policy and regulation. Wherever possible, the individual source of comments is noted in parentheses following the relevant text.[2]

[2]See http://www.rand.org/scitech/stpi/Evision/Transcripts/Day1-pm.pdf for a full transcript of this session, held on October 11, 2000. See http://www.rand.org/scitech/stpi/Evision/Supplement/ for the details of panel members' positions, as found in the full texts of their papers.

General Research Issues

Research is needed on energy-IT interactions.

The basic research challenge is to understand the ways in which the information technology revolution can decrease or increase energy use. (Reicher) Little investment in improving understanding of energy-IT interactions is reflected in DOE's current research (R&D) portfolio. (Moniz)

The government, however, is not that far behind the private sector. Utilities have just started looking at how IT affects their business processes. The IT industry has not yet perceived the energy sector as a significant customer sector. Nor have university-based researchers paid much attention to energy-IT interactions. (Lee)

Research is needed on the relationship between IT and energy demand—one of the most significant black boxes in our economy. (Jorgenson) For example, the portion of the economy affected by silicon chips has grown substantially, but the implications are not understood for energy systems. Research is needed on better characterizing the relationship of IT to improvements in productivity, reduction in energy intensity, changes in working habits, and resulting effects on energy use. (Lee) Energy intensity is the inverse of energy productivity.

Some IT panelists mentioned specific research questions, noting the fact that research is needed to even ask the right questions. (Allenby) For example, Jorgenson raised questions such as:

- How does IT substitute for energy in production inputs?
- How much of the slowdown in the growth of energy use since the IT investment boom began in 1995 is due to the substitution of IT for energy?
- Does energy use, like employment, fall with higher rates of IT investment?

The IT-based "revolution" in energy technologies offers opportunities to change thinking about energy systems (e.g. energy management, pricing, distributed generation, transmission and distribution). (Reicher) Targeted research is needed in both the energy technology and information technology areas, with a special emphasis on their interactions and dependencies. For example, IT allows design of feedback loop mechanisms into building systems. (McDonough) IT tools can also reduce the cost of energy to consumers. (Stram) Interactive design tools are needed to improve building design. (Christian)

IT's social dimensions affect energy systems.

Information technology is widely recognized as facilitating a major shift in production and distribution of products and services. The Internet is the latest in a long line of information-based revolutions (e.g. writing, printing, mail and telegraph, television) that have had a profound effect on societal processes. (Lee)

Social changes wrought by IT have implications for energy use, although few are well-understood at this time. For example, the energy implications of increased "telework" are ambiguous (Allenby). Distributed generation presents another example. Just as IT gave individuals access to information once limited to the elite, distributed generation will allow individuals (homeowners, businesses) to have more control over energy supply and use, and freedom from central control. (Richardson)

Demand Issues

Net effect of IT on energy consumption and productivity is unclear.

The IT-based economy has attributes that suggest an overall increase in energy demand on the one hand, and significant efficiency improvements on the other. The net result is unclear. The digital economy is based on the manufacture and use of semiconductors, a very energy intensive process. (Allenby) Power demands have increased in the digital economy (for example, through e-commerce) from 10 watts per square foot to over 100 watts per square foot in some buildings, stressing the energy infrastructure. (Arvizu) Alternatively, energy intensity is dropping faster than anticipated even as gross domestic product (GDP) increases.

The impact of the Internet/IT-based economy on electricity use is uncertain, but available evidence and credible analysis suggest that it is not having a significant impact. In fact, process efficiency is gained as information is substituted for energy in production processes. (Romm) For example, telecommuting has the potential to use less office space and reduces transport costs. Information use via the Internet/IT increases productivity without increasing other inputs (including energy). (Romm) The Internet can be used to optimize energy use by programming energy-using equipment (e.g., hotels can turn on lights/air conditioning in rooms at check-in, rather than running them all day). Home automation can accomplish similar energy savings. (Reed)

In determining the impact of IT on energy use, it is important to differentiate between the manufacture of an artifact, and the use and disposition of the artifact over its life cycle. (Allenby) For example, semiconductor manufacturing is energy- and water-intensive, but industry addresses these issues through process improvements, usually motivated by competitive pressure to lower production costs. The energy implications of IT-related equipment over their life cycle are not as well understood.

With IT-based business processes, there tends to be less inventory maintained, less floor space required, and less heating/cooling required. (Romm) One result is that IT may be changing the relationship of energy to economic growth, toward a type of growth that requires less energy-consuming activities.

IT facilitates different building design concepts. IT allows designers to improve efficiency by using less space for a given function, without necessarily compromising comfort or quality of life related issues. (Christian)

Supply Issues

Improved reliability and quality are emerging supply issues for an IT-based economy.

The IT-based new economy, dependent on the microprocessor, has increased the demand for highly reliable, high quality power. Premium power is being demanded on a megawatt scale. The demand for improved reliability and quality has stressed the existing infrastructure. "Presently, and for the next few years, the demand for power quality and reliability far exceeds that which has traditionally been provided by the electric power industry." (Arvizu)

Reliability is defined as the probability of disruption. The digital economy requires reliability of at least 99.999% (up from 99.9%). (Stahlkopf) Quality relates to constant voltage; minimal or no fluctuation is desired. Power fluctuations of 1/60th of a second can cause problems for online activities. (Stahlkopf)

The digital economy currently accounts for somewhere between 4 to 13 percent of electricity use; additional growth could eventually account for up to 30 percent of electricity use. Transmission and distribution systems will need major upgrades. More importantly, the design of the electricity grid and electricity generation systems needs to undergo significant rethinking, including the requirements for backup capacity.

Improved energy efficiency can help alleviate reliability problems by reducing demand. (Stahlkopf) Another possible response to the high reliability requirements of IT-based systems is to redesign digital systems to have less stringent requirements for electricity reliability and quality. (Pack) Identifying the optimum balance between redesigning energy-using equipment and upgrading energy infrastructure could be a productive research area. (Stahlkopf)

Distributed generation is another part of the solution to the reliability problem. (Stahlkopf) IT facilitates integration of distributed generation sources into the grid. (Friedericy) At the forefront of this trend are businesses closely connected to the Internet, who conduct significant e-business activity and are open 24 hours a/day, seven days a week. The cost of power failure for these businesses is "orders of magnitude greater than the cost of ensuring that the power doesn't fail. "(Arvizu) As a consequence, businesses are highly motivated to install the latest technology in on-site generation, power storage, and other equipment to control their own power supply.(Arvizu)

Silicon Valley companies assert that the cost of their products is insensitive to electricity price, but highly sensitive to power outages. (Stahlkopf) DOE has acknowledged there has been a lack of emphasis regarding electric system reliability in past R&D portfolios. (Reicher) This issue was raised two years ago in the first formal external review of DOE's R&D investment portfolio.

Policy and Regulatory Issues

IT allows real-time information on energy use and prices.

Real-time energy use measurement and intelligent monitoring are entirely new capabilities in the energy area. (Moniz) Consumers and businesses usually do not consider price when "buying" electricity because mechanisms to convey and track the information have been unavailable. IT communicates real-time information to consumers. For instance, smart appliances could convey electricity price information; consumers would then program use and temperature preferences to be sensitive to electricity price. Much of the underlying technology already exists to support real-time pricing. (Reed)

Real-time pricing raises several issues ripe for research, including: understanding of consumer response to this new information; articulating differences in application between residential and commercial sectors; scoping out appropriate government regulatory policy (e.g., standards for accuracy, certification processes); and estimating costs and benefits of achieving real-time pricing.

Other questions about real-time pricing arose at the conference:

- How would the interaction of real-time pricing and distributed generation work? (Lee)

- How would the implementation of real-time pricing likely affect infrastructure design? (Reed)
- What changes are needed to get real-time pricing adopted by both consumers and appliance/equipment manufacturers? Is there a possible role for federal legislation to encourage change? (Reicher)

The most recent external review of DOE's R&D portfolio identified several areas related to real-time information—real-time control and measurement, intelligent power, electronics, and smart metering—that require additional investment. (Reicher)

IT is having significant effects on deregulation and energy market structure.

Electricity deregulation has resulted in dramatic increases in the complexity of electricity infrastructure planning and has vastly increased the number of transactions over the grid. The transmission and distribution (T&D) system was not designed to support a deregulated generation market. Deregulation places an increased burden on T&D operators as the direction of "loop flows" (between buyers and sellers) changes frequently. The expansion of the T&D system has not kept up with economic growth and demand for power, largely because the incentives for industry investment in T&D are absent. The result is that the T&D system cannot support the reliability demanded by the digital economy. (Stahlkopf)

IT is facilitating energy industry restructuring and deregulation by allowing the processing of massive amounts of information. (Stram) For example, IT is providing the means for firms to outsource energy services and for suppliers and consumers to respond more efficiently to market signals. (Arvizu) IT can also improve the performance of the energy market by providing customers information on prices, thus enabling a demand-side response. (Reed)

IT enables real-time monitoring of system performance/status. (Allenby) This could change the way we manage buildings. (Romm) IT also facilitates the ability to respond remotely to systemwide needs. (Rosenfeld)

Several regulatory issues are associated with the increased prevalence of IT in the energy sector. For example, IT changes the analytical framework for standard setting (buildings, appliances) and trend analysis. (Reicher) The grid's inherent interconnectivity is similar to IT-based networks. Local disturbances on the grid cause systemic (global) problems. California's experience with poor electricity market performance after deregulation suggests that information is not flowing adequately. (Stahlkopf) Because the IT, telecommunications, and electricity industries are doing essentially the same thing—moving electrons—an integrated policy is needed to facilitate the eventual convergence of electric, natural gas, and IT/telecommunications industries. (McDonough, Richardson)

Worker Productivity, Building Design, and Energy Use

Issue

What indoor environmental factors affect employee productivity and to what extent would energy improvements in buildings improve worker productivity?

Panel Chair

Vivian Loftness
Chair, Department of Architecture, Carnegie Mellon University

Panelists

William J. Fisk
Staff scientist, Lawrence Berkeley National Laboratory, Berkeley, California

T. Keith Lawton
Director, Travel Forecasting, Department of Transportation, City of Portland, Oregon

David Wyon
Research Fellow and Adjunct Professor, Technical University of Denmark; Johnson Controls, Milwaukee, Wisconsin

This panel explored the relationship among building heating, ventilation, and air conditioning (HVAC) systems, worker productivity, and energy use. With the focus on buildings, most of the issues raised at the Policy Forum were on the demand side. Wherever possible, the individual source of comments is noted in parentheses following the relevant text.[3]

[3]See http://www.rand.org/scitech/stpi/Evision/Transcripts/Day2-am.pdf for a full transcript of this session, held on October 12, 2000. See http://www.rand.org/scitech/stpi/Evision/Supplement/ for the details of panel members' positions, as found in the full texts of their papers.

Demand Issues

The study of worker health and productivity, the role of building design in worker health and productivity, and implications for energy use requires an interdisciplinary research program.

Panelists suggested that the lack of clear definitions of worker productivity has led to confusion about the underlying relationships among worker health, productivity, and the "built" environment, referring to conditions inside buildings. There has also been difficulty in determining the applicability of research in one domain (e.g., sick building syndrome) to other domains (e.g., energy use in modern commercial office space). (Wyon, Rosenfeld, Loftness).

Panelists argued that health is a critical part of worker productivity and, further, that worker productivity is related to energy consumption. They discussed opportunities for improving health and productivity by providing better environments in office buildings. For example, Fisk provided estimates of the economic benefits accruing from reductions in respiratory illnesses that may be triggered by inadequate building ventilation and heating. Some evidence was presented suggesting that changes in building systems and design may increase worker performance while also decreasing energy use. For example, changes in heating, ventilation, and air conditioning (HVAC) systems may improve worker health while also reducing energy consumption.

Panelists suggested that EERE could play a decisive role in initiating and coordinating a research agenda in these areas. Some attendees suggested the creation of a virtual institute, supporting multiple avenues of interdisciplinary research under the umbrella of the EERE. For example, EERE could:

- support the study of the relationships between worker productivity and the built environments and their impacts on energy use
- act as a clearinghouse of information on the research findings from various disciplines to encourage spillovers in knowledge to other disciplines
- provide demonstration sites of buildings or parts of buildings to test research results and to educate researchers, architects, and building developers on the economics of incorporating building design changes.

The last point would generate data that might be used to justify potentially higher up-front building costs with improved worker productivity and lower energy costs.

Innovation in building design requires a systems approach, possibly spanning multiple federal agencies.

By current estimates, buildings use over one-third of all energy in the U.S., principally for cooling, heating, and lighting. (Loftness, Wright) Further, buildings are significant sources of water use, wood harvest, and raw material use. As noted above, various panelists and conference attendees were concerned about the adverse impact of modern commercial buildings on occupant health, in turn affecting worker performance and productivity. A systems approach integrates each of these factors into a coherent framework to guide building design.

Panelists and speakers voiced a need for a new type of researcher educated in "design science," "design engineering," and "design economics." (Loftness, McDonough, Christian) A curriculum focused on design science would enable researchers to consider the relationships between worker productivity and health in a built environment. Education in design engineering would provide the capabilities needed to design effective systems for buildings. Courses in design economics would provide expertise in life cycle decisionmaking, incorporating up-front design decisions with long term costs and benefits of those choices.

Conference attendees encouraged EERE to look at the gains accruing from healthy built environments, to consider safety and security issues, and to look at the educational potential of buildings. McDonough and others suggested that, through R&D, the U.S. could become an exporter of advanced building systems and design technology to the rest of the world. However, there was a pressing need for investment in research to advance new building components and building construction methods. Researchers called for support from the DOE, including (1) federal funding to encourage product research in the building industry; (2) support for interdisciplinary and interagency approaches to understanding the built environment, involving health specialists, air quality specialists, environmental specialists, statistical analysts; and (3) support for Ph.D. students in building research. (Rosenfeld, Loftness, Wyon).

Little research has been conducted on the relationships between land use planning and energy use.

Cities in the U.S. have evolved over the last 50 years with little forethought or planning. As a consequence, older cities are not energy-efficient and are difficult or impossible to retrofit. (McDonough, Lawton) Several speakers focused on the importance of taking a systems approach to understand the relationships between the built environment, quality of life, and energy use. (Lawton, McDonough, Kampshroer, Allenby) For example, in a study of travel times devoted to work and nonwork activities, Lawton provided evidence of increasing vehicle miles per day per person in the U.S. due to changes in housing choices (e.g., suburban vs. urban), job locations (e.g., technology parks located outside cities), and accessibility to activity spaces.

Panelists suggested that EERE might consider

- directing and supporting research on the relationships between land use planning and energy use; on land use patterns and quality of life; and on the transportation infrastructure, land use, and energy
- providing forums to foster information exchange with city planners
- playing an technical advisory (nonregulatory) role in land use discussions.

Little research exists on the impact on energy use of increased incidence of working at home.

The rapid diffusion of information and communications technologies has increased the ability of people to work out of their homes. Conference attendees had varying opinions about the impact of this trend on energy consumption. At-home workers may relieve space demands in office buildings, reduce building size requirements and associated energy use, and eliminate energy associated with commuting. (Romm) However, residential housing may often be less energy-efficient than modern office buildings, potentially increasing systemwide energy costs. Accordingly, while information technologies allow work at home, little is known about the impact of teleworking on energy systems. Support from the EERE might include research on

- internal environmental quality of residential and commercial buildings
- energy-efficient systems for residential housing
- incentives for at-home workers to adopt energy-efficient practices and systems.

Research on how consumers make decisions needs to be extended to how consumers make decisions about energy use in their homes and in commercial buildings.

Technology currently allows consumers to monitor and remotely control various electrical devices in their homes. Conference participants debated the likelihood of consumers, particularly in residential settings, to use this and future monitoring technology effectively to make decisions on energy use. In particular, participants were concerned that there would be no change in energy use behavior when consumers were provided with real-time energy cost data. Commercial building managers, however, were suggested to have the most to gain from these technologies and, indeed, some already were making use of available technologies.

In order to understand the value of developing technologies that would allow consumers to monitor real-time energy costs, EERE might support behavioral research on how people make decisions about energy use in their homes and in commercial buildings.

Supply Issues

There are opportunities for using research on buildings and building design to advance and diffuse new energy generation technologies.

Some conference attendees suggested that energy technology, such as fuel cells and biomass-fueled generators, could be tested and advanced through trials in commercial and public buildings. Coupled with this, attendees also commented on the need for research on distributed energy technology for buildings to foster its diffusion and possibly provide energy back to the electrical grid.

Potential areas of EERE involvement include

- providing demonstration sites of buildings or parts of buildings that are powered by alternative energy sources to help educate researchers, architects, and building developers on technology developments
- fostering research collaborations in energy generation and building design
- providing tax incentives to encourage the adoption and diffusion of energy supply technology in building design
- disseminating information on distributed generation technology and evolving energy generation systems.

Systems Approach to Energy Use

Issue

What technology and design changes could tap the savings potential and productivity enhancements of more systems-based energy generation and use?

Panel Chair

Maxine Savitz
General Manager, Technology Partnerships, Honeywell International

Panelists

Rose McKinney–James
Vice President, Government Affairs, Faiss Foley Merica

Josephine S. Cooper
President and CEO, Alliance of Automobile Manufacturers

Jack Cogen
President, Natsource

Charles Ebinger
Vice President, International Energy Services, Nexant

Robert Sonderegger
Director of Modeling and Simulation, Silicon Energy Corporation

This panel examined the role and benefits that systems approaches to energy development and use could potentially yield. Panelists and other participants discussed several different kinds of systems approaches: restructuring of the electric utility industry; promotion of methods of encouraging consumer adoption of new technologies; use of emissions trading to stimulate investment in new energy technologies while lowering emissions of pollutants and greenhouse gases; and complementarity among different kinds of energy sources and generation technologies including distributed generation and renewable energy. Observations made during this panel discussion are organized around these topics. Wherever possible, the individual source of comments is noted in parentheses following the relevant text.[4]

Changing Patterns in the Energy Industry

Domestically and globally, the institutions, markets, and technologies that characterize the energy industry are undergoing massive change.

The domestic and global energy industries are being deregulated, unbundled, and restructured. This has led to major increases in nonutility electric power generation, repowering of existing generation facilities, and a reexamination of whether aging nuclear plants should be retired or granted life extensions. As part of deregulation, rapidly growing spot, forward, and futures markets have emerged for electricity, natural gas, and emissions allowances. And while these markets have seen oil and gas prices rise and become increasingly volatile, coal prices have remained low and stable. Finally, new distributed generation technologies and progress in developing renewable energy sources have the potential to change the nature of electric power generation. (Ebinger)

Fossil fuels need to be used more efficiently throughout the energy chain including during extraction, conversion, and end use, in parallel with continued advances of new nonfossil technologies.

Global population growth, rising fossil fuel consumption and greenhouse gas emissions, and accelerating urbanization are creating many energy-related environmental problems worldwide, especially in the major cities of the developing world, where the bulk of the world's future population growth will occur. (Ebinger)

In general, we need to make dramatically more efficient use of traditional fuels, which are mostly fossil based. That means increasing generation efficiency and end-use energy efficiency. (Reicher)

---

[4]See http://www.rand.org/scitech/stpi/Evision/Transcripts/Day2-pm.pdf for a full transcript of this session, held on October 12, 2000. See http://www.rand.org/scitech/stpi/Evision/Supplement/ for the details of panel members' positions, as found in the full texts of their papers.

Role of Renewable Energy and Demand-Side Management

Renewable energy and demand side management can provide near-term solutions to phased deregulation while reducing risk and uncertainty.

California's failure to deregulate its electricity markets quickly has contributed to a mismatch between demand and supply. As economic growth continues to increase demand, California's problems may be repeated in other states that deregulate slowly—unless new solutions to these shortages can be found. This is because the traditional solution to shortages is to build more supply. Unfortunately, traditional supply technologies take years to build and they increase exposure to fuel price volatility as well as increased emissions-related liabilities.

Renewable energy units and demand-side management can be brought online relatively quickly since they tend to be smaller and more scalable than conventional plants. Because they do not require fossil-fuel inputs, renewables provide a long-term strategic way to reduce exposure to fuel price volatility. They also avoid the liabilities associated with the onset of new and more aggressive air emissions regulations aimed at traditional pollutants, greenhouse gases, or other constituents. (Cogen and Varilek)

Renewable energy and demand side management have public benefits that are not reflected in market prices. This results in underinvestment in these technologies.

In addition to closing the supply-demand gap and reducing exposures to fuel price volatility and potential future emissions costs, renewable energy and demand-side management provide additional public benefits. For example, these technologies increase national energy independence and improve air quality. Unfortunately, since these benefits are not reflected in existing market prices, these technologies are not deployed at socially optimal levels. Part of the reason for this underinvestment is that some of those who enjoy the benefits of these technologies do not have to pay for them. (Cogen and Varilek)

Renewable energy and demand side management can lower conventional and greenhouse gas emissions as well as the cost of those emissions reductions.

Integrating renewable energy and demand-side management into emissions trading markets could reduce emissions while lowering the cost of those reductions. For example, these technologies could be integrated into today's sulfur dioxide and nitrogen oxide programs. And as the small but growing precompliance market for greenhouse gas emissions reveals, they could play a major role in reducing greenhouse gas emissions and lowering total costs should an international agreement emerge in the future. (Cogen and Varilek)

A more sustainable approach to structural design would be to design toward a goal of minimizing inefficient use of or eliminating nonrenewable energy needs over time.

This design approach would use the three strategies of energy conservation, energy efficiency, and renewables energy. Acting together as a triad, such a "zero net energy" approach would create awareness of the importance of sustainability, how it can be achieved, and that design and consumer choices can make a difference. Energy conservation, energy efficiency, and renewable energy should become the foundation of a more sustainable "zero net energy" approach to design. (McKinney-James)

Designers, builders/manufacturers, suppliers, and users must work together as members of a strategic partnership. Local utilities, state agencies, officials, and home builders must be forged into a strategic partnership. (McKinney-James)

Other Demand Issues

Creating demand for energy-efficient products requires a simple message and an iterative and transparent process.

The need to create demand and facilitate the introduction of new technologies was confirmed by the automotive industry, who argued that change needs to be seamless or transparent to be accepted by consumers. (Cooper)

People like choice, but they tend to prefer small evolutionary changes that are easy to compare to the status quo. Renewable energy systems must be easy to install and they must pass the tests of reliability, aesthetics, and performance. The federal government has generally focused on developing new energy technologies rather than promoting their use.

Increasing energy efficiency can be achieved throughout the entire energy system.

Two of the fastest growing markets in the developing world are India and China. Home appliances such as air conditioners and refrigerators are growing at 20 to 30 percent per year. Transportation is also growing quickly. It is very important that these markets be filled with energy efficient appliances and transportation technologies because the alternative of millions of people using inefficient technology would be staggering. (Ebinger)

Most new transportation technologies cannot compete on a cost and power-per-unit volume with gasoline.

During the last century, the industry significantly improved the internal combustion engine by looking at the vehicles, engines, and fuels as a system. Today, the industry is looking at a large number of new technologies, some of which promise further increases in efficiency and reduced emissions. The next generation of technologies includes advanced gasoline and diesel engines, alternative-fueled vehicles, battery and hybrid vehicles, and fuel cells. However, most of these engine/fuel systems cannot yet compete on a cost and power-per-unit volume with gasoline. (Cooper)

New engine/fuel systems may require new fuels and new fuel infrastructures that will be expensive and possibly unpopular with consumers.

Strategic introduction of new technologies such as fuel cells into residences first could help introduce the technology and the fueling infrastructure to people gradually. As R&D on vehicular use of fuel cells continued, citizens would be more likely to accept the technology when it is finally transferred to vehicles. (This approach could potentially move a vehicle's fueling infrastructure into one's garage while also helping to improve fuel cell performance and lower their costs through larger production volumes.) (Cooper)

Information technology could improve energy and environmental performance of automobiles.

Information technology could improve the energy efficiency, environmental performance, and general mobility of individuals and their vehicles by optimizing operations within the vehicle as well as on the roads themselves. It has also made it possible to trade electricity, digitize customer services, improve procurement speed, streamline supply chain management, and reduce costs. (Cooper)

Supply Issues

The nation needs to ensure that affordable, adequate, and environmentally sustainable sources of energy can meet domestic and global energy needs.

R&D policy and budget priorities should be guided by long term goals, as well as the potential domestic and international export markets for various energy technologies. Each of these is fundamentally affected by the nation's approach to addressing climate change. Internationally, 33% of the global increase in primary energy use of over the next 20 years is expected to occur in India and China alone. Similarly, these two countries alone are expected to account for 97% of the world's increased use of coal. Seriously addressing climate change would suggest that the U.S. work with these countries to develop alternatives to coal or to at least help them to burn coal more efficiently. (Ebinger)

Technical solutions to our energy supply needs should include increased efficiency, higher efficiency in production productivity, clean coal technologies, renewable energy, hydropower, and nuclear energy.

To keep natural gas prices affordable, coal gasification technologies should be considered. Additionally, investing in R&D on clean coal technologies could yield many benefits within the U.S. and internationally. (Ebinger)

Renewable energy resources could play a significant role as R&D continues to improve their performance and lower costs. These technologies could be particularly important in developing countries, if an international agreement on reducing greenhouse gas emissions is adopted. (Ebinger)

Hydropower and nuclear power could theoretically make a resurgence. Hydropower, if adapted to reduce environmental impacts through new technologies (e.g., fish-friendly turbines), could make hydro more acceptable. Nuclear power has the technical potential to address energy supply needs, but to be viable, the nuclear waste problem must be solved and the technology's public acceptance problems be addressed—a task which many see as nearly insurmountable. (Ebinger)

The value of hedging strategies is becoming apparent.

Some energy companies are beginning to build hybrid facilities that combine the best of natural gas generation (its reliability) with the best of wind (its low cost). The wind helps hedge against the volatility of natural gas prices while natural gas reduces the intermittency problem of wind. Mixing these technologies, together with energy efficiency, can lead to dramatic improvements in generation and end-use efficiency if stimulated by proper policies and market structures. (Reicher)

Distributed Generation

Electricity deregulation is fundamentally changing the electric power industry, including creating opportunities for distributed generation (DG) and increased use of cogeneration.

Electricity deregulation has begun to break down the traditional roles of utilities as energy suppliers and consumers as energy users. One emerging example of these changes is the opportunity to decentralize generation at a large number of distributed locations spread out over the grid. While distributed backup generation and cogeneration have been used for a long time, new schemes are emerging that would resell power from traditional backup generators and cogenerators into secondary markets. This would make power and heat available to facilities connected to the grid or in close proximity. (Sonderegger)

Today's control technologies and management approaches are inadequate for maintaining reliability and optimizing operations in a fully distributed system.

Evolving the grid from a central generation model toward a distributed generation model poses significant challenges to the current Supervisory Control and Data Acquisition (SCADA) techniques used to manage the electric power grid. As more actors produce and consume power simultaneously, the number of interconnections increases, making managing all possible combinations of power production and consumption for reliability, much less optimizing operations, much more complex. The increased complexity occurs because dispatching generation resources at the right times and locations requires reliable communication networks and fail-safe monitoring and control equipment both for distributed generators and loads. (Sonderegger)

While the Internet may appear to provide the modern solution to these problems, several significant problems remain. First, because the Internet can experience congestion or outages, fail-safe control strategies need to be developed that "do no harm" in the event of communication problems as well as changes in generation, demand, and transmission and distribution. Second, the number of communication and control schemes for distributed resources is already proliferating, posing a significant barrier to having one system that communicates with all of them. Third, performing complete or even partial optimizations in real time can be extremely complex and daunting. (Sonderegger)

Implementing distributed generation will require new technologies and techniques to ensure reliability and optimize operation.

Near-, medium-, and long-term solutions are needed because no perfect solutions currently exist. Internet-based computer systems with extensive safety and fail-safe measures will be needed. The long-term solution to this problem will likely use an Internet-based system of computer-automated auctions that match the bids and requests filed by intelligent computer automated generation agents, load agents, transmission agents, and energy storage agents. (Sonderegger)

In the nearer term, a three-tiered control scheme that is compatible with existing control techniques could help the grid take the first steps toward decentralization. The three tiers would be at the device, site, and regional grid operator level. (Sonderegger)

Firms are already installing distributed generation, but regulatory and technical obstacles cause these systems to be isolated from the grid.

Regulatory and technical barriers have led many firms installing DG to configure them as grid-isolated systems, rather than parallel systems. This is unfortunate—particularly in supply- constrained areas, such as California—because isolated systems prevent the generation capacity and ancillary services from being available to the grid. Grid-isolated systems forfeit the benefits that DG offers to society at large and prevent the owners of those generation assets from profiting from power sales made to the grid. (Sonderegger)

Uniform technical requirements, fair pricing, and open participation in competitive markets will allow DG to contribute to meeting our electric power needs.

DG can begin contributing toward meeting our electric power needs and beginning the evolution toward distributed generation if: (1) uniform requirements are set under which DG is permitted to supply to the grid; (2) market mechanisms are created that fairly price DG power according to its value to the utility (for example, according to avoided transmission costs); and (3) existing Independent System Operator (ISO) and Power Exchange (PX) markets are extended to include an active market of DG buyers and producers within utility service territories and across different utilities. (Sonderegger)

Emissions Trading

Market-based allowance trading systems let the regulated parties decide the most cost-effective way to use available resources to comply with emissions requirements.

Traditionally, air pollution has been regulated by a command and control approach where the federal government mandated the use of a specific emissions control technology. This approach forced businesses to buy the technology, even if there were more cost-effective ways to reduce those pollutants both at the plant-level as well as nationally.

The 1990 Clean Air Amendments replaced this command and control approach with a completely new and innovative market-based emissions allowance trading system to limit emissions of sulfur dioxide. Under this system, a firm must own one sulfur dioxide allowance for each ton of sulfur dioxide the firm emits over a given year. Firms are allowed to buy and sell allowances among themselves. Thus, firms that can most efficiently reduce their own emissions may choose to sell excess allowances to firms with higher marginal costs to achieve emissions reductions. By controlling the total number of allowances, the Environmental Protection Agency (EPA) can control the aggregate production of sulfur dioxide. Allowance trading creates incentives for new technologies and processes that lower the cost of compliance for individual firms as well as the nation. (Cogen and Varilek)

Careful design of emissions markets is crucial to creating proper incentives for investing in these technologies.

The emissions reduction and cost reduction opportunities of these technologies will only be realized if the problems in today's existing markets are corrected, and if new or emerging markets are designed correctly from the start. (Cogen and Varilek)

The general deficiency in today's emissions trading programs is that different investors receive different rewards for the same investment in renewable energy generation and demand-side management. This leads to underinvestment and lost opportunities for cost-effective emissions reductions. The discrepancy occurs because investment incentives exist for owners of regulated emissions sources, but not for independent investors who are not affected by the emissions trading restrictions. The key difference between these groups is that the former can use investments in renewable energy and demand-side management to balance actual emissions and permit holdings, while the later cannot. (Cogen and Varilek)

The first successful emissions reducing mechanism will likely establish the international standard.

Following the establishment of the Kyoto Protocol in 1997, serious efforts began to determine what types of mechanisms could be used to reduce emissions globally. Today, several countries and even a few firms have adopted or experimented with different greenhouse gas reducing regimes. In addition, a small international precompliance market has already emerged. The first of these mechanisms to succeed may set an international precedent and guide international negotiators toward such a system even if it is less than perfect. (Cogen)

Thus far, the best emissions trading system seems to be a cap-and-trade system that uses finite, high-integrity allowances, and that can issue allowances for deploying renewable energy generation and demand side management. (Cogen and Varilek)

Markets will reveal the anticipated marginal cost of carbon abatement well into the future. These prices will provide both the public and private sectors with insight on what near-, medium-, and long-term R&D they should pursue. (Cogen)

Regulatory and Policy Issues

There are numerous ways to affect the national energy system, but they are difficult to coordinate.

The nation has many policy, regulatory, and financial tools available to affect the nation's energy system directly. These include environmental regulations, electricity deregulation, the tax code, international trade, and the nation's rural and agricultural policies, as well as the nation's investments in energy R&D. (Rattien, Savitz, Reicher)

The federal government can do many things to accelerate the demonstration and deployment of new technologies.

For the past several decades, federally funded work on energy technologies has largely focused on technology development and little attention has been given to markets. (McKinney-James)

There are several things that the federal government can do to help introduce new technologies. These include: passing national electricity legislation; refocusing R&D to address market forces; improving demonstration programs, information dissemination, and advocacy; using the federal government to drive markets and create demand (by lowering federal procurement barriers); and new financing tools that take life cycle costs into account.

Comprehensive national electricity legislation is needed.

Perhaps the most important thing that needs to be done is for Congress to pass a well-designed, market-oriented comprehensive national electricity policy. This will end the state-by-state deregulation activities that are currently underway, keeping investment decisions uncertain and making it difficult and expensive to introduce new technologies. (Ebinger)

Legislation must be comprehensive so that a uniform and fair policy on interconnection is set. This is crucial because it will allow the private sector to focus on making good products, rather than marketing those products in each of the 50 states. Setting national standards will help promote the deployment of new technologies such as distributed and renewable energy generation. Creating a clear national set of rules that treat renewable energy appropriately will also help the private sector see green power as a business opportunity that is worth investing in. (Ebinger, Savitz, McKinney-James)

Demonstration Programs Are Needed.

Many of the new technologies that have been developed are being constrained not by a lack of capability, but because their performance has not been demonstrated to consumers, and they have not been able to reduce their costs due to the low volume of production. Until these new products are demonstrated and their costs are reduced, these new technologies will not penetrate the market. (Savitz)

The federal government can also help through a variety of other means including information. The Energy Star program for example, makes it simple for people to buy efficient equipment and appliances. Energy Star is a voluntary labeling program designed to identify and promote energy efficient products and reduce carbon dioxide emissions. (Reicher, McKinney-James)

State and local governments can help shift national policy while bringing local benefits.

State and local governments can also affect energy supply and energy use in significant ways. Examples include renewable portfolio standards, systems benefit charges, highly visible demonstration/deployment programs, as well as a well-identified central repository for renewable data. Each of these can help overcome technical issues and create public and consumer awareness. (McKinney-James)

Open Forum

Moderator

Dan Reicher
Assistant Secretary for Energy Efficiency and Renewable Energy,
U.S. Department of Energy

The third morning of the conference was devoted to an open forum for conference attendees who had not previously had an opportunity to present their views.[5]

Speakers

James Winebrake
Associate Professor, James Madison University

Winebrake posed the question of whether EERE has the mechanisms in place to learn from yesterday, to identify their preferred futures, and to initiate the programs to achieve those futures. EERE needs to think proactively about what is going to happen in 10, 20, or even 50 years. It needs to develop some informed strategic planning to help achieve that desirable future, and then track key indicators, to help guide program planning and policy direction. Finally, EERE needs some coordinated research. He suggested a research institute that would perform these and other functions such as technology and policy assessment, design practical decision analysis tools that can be used by DOE decision makers to help them incorporate the desired future into their present day planning, and work with the state energy offices to get them thinking about the future.

Kevin Kephart
Director, Agricultural Experiment Station, South Dakota State University

DOE should organize support for agricultural schools who are researching environmental quality in buildings, carbon sequestration and biomass energy

[5]See http://www.rand.org/scitech/stpi/Evision/Transcript/Day3-forum.pdf for a full transcript of this open forum, held October 13, 2000.

sources. Biomass sources of fuel for transportation could boost the agricultural sector of the economy.

Stefano Ratti
Assistant Director, Center for Environmental Energy Engineering, University of Maryland

Ratti spoke about the benefits of on-site generation of electricity from natural gas along with utilization of thermal energy available from the power generation process. Buildings account for almost 40% of U.S. energy consumption and they have a major impact on national energy use. At the very moment, the bulk of that energy is generated by large power plants in a traditionally inefficient process that wastes large amounts of thermal energy and releases an enormous amount of carbon dioxide. The thermal energy that otherwise is wasted could be utilized for heating and cooling. The most promising emerging technologies in power generation are fuel cells and microturbines that can be effectively integrated with advanced cooling technologies. Convincing demonstrations are needed to provide insight into the optimal integration of these new technologies for commercial buildings.

Darien Unger
Researcher, Energy Laboratory, Massachusetts Institute of Technology

Unger described how gas turbines have become predominant in the power generation field, taking about 70% of orders and market share away from the traditional steam turbine methods. Gas turbines took over the market for four reasons: technological improvement in terms of new alloys; fuel availability and affordability; environmental regulations that raised the cost of competing sources of power and energy technologies; and electric restructuring, mainly on the state level, that allowed the introduction of competitive pressures in the energy and power markets. State level restructuring also encouraged the introduction of other factors like information technology that facilitate competition.

Jerry Asher
Associate Director, Electric Vehicle Association of Greater D.C.

Education is a key component of building a vision of the future. As an example, DOE should sponsor electric vehicle competitions in high school and college levels.

Jeffrey Christian
Director, Buildings Technology Center, Oak Ridge National Laboratory

Christian suggested the establishment of a profession called building science that would create the intellectual framework for integrating multiple objectives for building design.

Michael Markels
President and CEO, Ocean Farming Inc.

Markels discussed the need for a technology to reverse the increase of carbon dioxide in the atmosphere. He believes that the ocean could be used as a repository for large amounts of carbon in the form of biomass and carbon dioxide. DOE should support ocean-based carbon sequestration programs.

Hans Friedericy
Consultant, Honeywell Aerospace

DOE should conduct research on hydrogen technology/fuel cells that show promise for replacing batteries in low-power applications. He also spoke about his company's vision of hydrogen energy for the United States and eventually for the world. He believes that industrial applications of distributed generation, improved grid management, and ultraclean fuels hold promise for reliable and sustainable power.

Megan Smith
Codirector, Biofuels Transportation Issues, American Bioenergy Association

DOE should consider additional support for research on biomass as a source of fuel for transportation.

Jeffrey Harris
Group Leader, Government and Industry Programs, Lawrence Berkeley National Laboratory

DOE should sponsor more social science research on how individual decisionmaking affects energy choices. For example, Washington State has done some work on understanding how government organizations in particular make purchasing decisions related to energy efficiency. Research from the behavioral sciences should complement traditional R&D and product development to enhance market transformation.

Kevin Dye
Chief Process Scientist, CWA Ltd.

DOE should support research and teaching of energy systems and design concepts. While there are calls for systems-based approaches within government and the private sector, most every systems science department in U.S. universities has gone under over the last 15 years.

Arthur Rosenfeld
Commissioner, California Energy Commission

DOE should consider support for an interagency institute for indoor environmental quality. This kind of institution could generate the quantitative infrastructure that building engineers need for design change.

Mark Berman
Principal, Business Development, Davis Energy Group

R&D priorities might include advanced microturbines, fuel cells, photovoltaics, reformers for hydrogen, and biomass. A second priority is product development: how to take proven technologies and apply them in a way that brings the benefits to as many people as possible, saving some energy, enhancing performance of buildings, and worker productivity. A third priority would be the creation of a virtual research institute for high-performance buildings. Builders need tangible evidence to show bankers, lenders, and investors.

Scenario Analysis

Richard Silberglitt
RAND

Anders Hove
RAND

RAND undertook an analysis of future scenarios to help inform EERE's planning process.[6] Scenarios are used as descriptions of alternative futures, not as forecasts or predictions. They enable policymakers to systematically consider uncertainties inherent in energy planning and to select strategies that are robust. Robust strategies perform adequately over a range of conditions, in contrast to those that do very well under some conditions, but fail under others.

This section summarizes the key points and findings of the scenario analysis undertaken as part of the E-Vision 2000 process. The purpose of this summary is to indicate the range of representative scenarios that are documented and familiar to energy experts, aggregate these scenarios into a smaller set that can be clearly distinguished from one another, and illustrate some of the strategies implied by these scenarios. It is important to note that this analysis did not extend into a full-fledged strategic planning process.

In the planning phase of the E-Vision 2000 process, the scenario analysis was anticipated to provide conference participants with a common basis for discussion of widely available energy scenarios. In practice, the insights from the scenario analysis were not integrated into the panel discussions of the Policy Forum because the agenda lacked sufficient time. By design, they were not used by participants in the Delphi process either.

The analysis did highlight essential similarities and differences among the many energy scenarios that have been published in recent years, and served to illustrate how such scenarios and studies can provide policy insights despite the uncertainty associated with long-term projections. Additional work is needed to take these scenarios to the point where the implications for energy R&D can be more clearly focused and useful to EERE.

---

[6]See http://www.rand.org/scitech/stpi/Evision/Supplement/scenario.pdf for the full text of RAND's scenario analysis and the accompanying graphics.

Method of Scenario Analysis

RAND reviewed 18 scenarios describing energy consumption and fuel mix over the next 20 years or so, as summarized in Table 1. These scenarios were representative of a broad range of institutional and individual perspectives, were readily available within the time constraints of preparation for the E-Vision 2000 project, and had sufficient detail to allow analysis. It is not an exhaustive list, but it includes most of the major energy scenarios in common use within the energy community.

An initial review clearly indicated significant overlap among many of the scenarios. Rather than dealing with each scenario separately, RAND's scenario analysis team sought a means of grouping similar scenarios together in clearly defined clusters. This "meta-scenario" approach would provide a common framework to compare and contrast the scenarios listed in Table 1 in meaningful terms for policy development.

These meta-scenarios were then assessed to determine the illustrative signposts, hedging strategies, and shaping strategies needed for an adaptive approach to strategic energy planning.[7] *Signposts* alert policymakers to the approach of undesirable futures or gauge our progress toward desirable futures. *Hedging strategies* are robust actions that work well under various uncertain futures. *Shaping strategies* are actions, intentional or otherwise, that can alter the course of future events, and help policymakers achieve desirable futures. These elements—signposts, hedging strategies, and shaping strategies—are the building blocks of an adaptive approach to strategic planning that can help the U.S. avoid undesirable outcomes and move toward desired futures despite the uncertainty inherent in energy planning.

Framework for Analysis

A set of parameters was defined that could provide a common framework to compare and contrast the large number of commonly used energy scenarios in ways that are meaningful for policy development. This framework was then used to identify groupings of individual scenarios that constituted "meta-scenarios," representing a plausible set of alternative futures. Parameters were defined to provide a common framework to compare and contrast scenarios.

[7]James A. Dewar, Carl H. Builder, William H. Hix, and Morlie H. Levin, *Assumption-Based Planning: A Planning Tool for Very Uncertain Times*, Santa Monica, Calif.: RAND, MR-114-A, 1993.

Table 1

Models and Scenarios RAND Compiled and Evaluated

Model/Scenario	Source	Notes
EIA	*Annual Energy Outlook 2000* (EIA, 2000)	The Energy Information Administration (EIA) Reference Case has five variants, and 32 side cases (20 of which were fully quantified). These scenarios are extrapolations of current trends and policies, using a combination of econometric and technological models.
EIA Kyoto Protocol 1	*Impacts of the Kyoto Protocol on U.S. Energy Markets and Economic Activity* (EIA, October 1998)	This includes six scenario variants that use the EIA economic and technological models, with an added carbon price component included in the price of each fuel. The report also describes five sensitivity cases that vary economic growth, rate of technological improvement, and nuclear power use.
EIA Kyoto Protocol 2	*Analysis of the Impacts of an Early Start for Compliance with the Kyoto Protocol* (July 1999)	The second of EIA's two analyses of the impacts of Kyoto, revisited the same assumptions together with implementation beginning in 2000. The carbon prices were reduced somewhat but the conclusions were unchanged.
Econometric Scenarios	International Energy Agency (IEA); Gas Research Institute (GRI); American Gas Association (AGA); Independent Petroleum Association of America (IPAA); Standard and Poors' DRI Division; Wharton Econometric Forecasting Association (WEFA); Source: EIA, Annual Energy Outlook 2000 (EIA, 2000)	Scenarios based upon econometric models developed by multinational and nongovernmental organizations were included in the study.
WEC	*Global Energy Perspectives* (Cambridge University Press, 1998)	The World Energy Council (WEC) and the International Institute for Applied Systems Analysis (IIASA) describe 6 world energy scenario variants that span a broad range of alternative futures.

Table 1—Continued

Model/Scenario	Source	Notes
Royal Dutch Shell	Royal Dutch Shell Energy Group, available online at www.shell.com, accessed July, 2000.	Royal Dutch Shell developed one scenario variant in which growth in energy consumption is sustained at a high rate, and another variant in which "dematerialization" slows energy consumption.
IPCC	The Intergovernmental Panel on Climate Change (IPCC), *The Preliminary SRES Emissions Scenarios* (January 1999)	IPCC described six scenario variants with different assumptions about economic, population, and technological growth.
America's Energy Future	The American Council for an Energy Efficient Economy (ACEEE), Alliance to Save Energy, National Resource's Defense Council, and the Union of Concerned Scientists, in consultation with the Tellus Institute, *America's Energy Future*, (1997)	ACEEE described three scenario variants based upon high energy efficiency and investment in renewable energy, together with substantial changes in the energy infrastructure.
Bending the Curve	Stockholm Environment Institute and Global Scenario Group, *Conventional Worlds: Technical Description of Bending the Curve Scenarios* (1998)	The Stockholm Environment Institute and Global Scenario Group describe two scenario variants driven by intervention to reduce carbon emissions and transition to renewable energy sources.
Interlab Working Group 1	Inter-Laboratory Working Group, *Scenarios of U.S. Carbon Reduction: Potential Impacts of Energy-Efficient and Low-Carbon Technologies by 2010 and Beyond* (1997)	First of two reports by the five DOE national laboratories describes two scenario variants in which public policy actions and market intervention lead to reduced carbon emissions.
Interlab Working Group 2	Inter-Laboratory Working Group, *Scenarios for a Clean Energy Future* (2000)	Second of two reports by the five DOE national laboratories; describes three scenario variants involving public policy actions and market interventions designed to bring about reduced carbon emissions.
PCAST	President's Council of Advisors on Science and Technology (PCAST), *Powerful Partnerships: The Federal Role in International Cooperation on Energy Innovation* (June 1999)	PCAST makes quantitative estimates of reductions in fossil fuel use, U.S. oil imports, and CO_2 and other emissions possible with increased investment in energy RD&D.

Table 1—Continued

Model/Scenario	Source	Notes
Ausubel	Jesse Ausubel, "Where is Energy Going?" (*The Industrial Physicist*, February 2000)	Ausubel of Rockefeller University describes the decarbonization of the fuel mix in "pulses" of rising energy consumption per capita, with natural gas as the 21st century transition fuel to hydrogen.
Romm et al.	Joseph Romm, Arthur Rosenfeld, and Susan Herrman, *The Internet and Global Warming* (1999)	Romm et al. argue that e-commerce spurred recent improvements in U.S. energy efficiency, and posit future increases in efficiency beyond extrapolation of current trends, with concomitant reductions in energy consumption.
Lovins and Williams	Amory Lovins and Brett Williams, *A Strategy for the Hydrogen Transition* (April 1999)	Lovins and Williams envision stationary fuel cells powering buildings and providing distributed generation of electricity, resulting in the reduction of size and cost of fuel cells and hydrogen infrastructure, and ultimately cost-effective fuel-cell-powered ultra-high efficiency "hypercars."
CARB	California Air Resources Board (CARB), *Status and Prospects of Fuel Cells as Automobile Engines* (July 1998)	CARB examined the cost of hydrogen infrastructure for automotive fuel cells, as well as methanol and gasoline as hydrogen sources.
ADL	Arthur D. Little (ADL), *Distributed Generation: Understanding the Economics* (1999)	ADL provides a detailed market study of fuel cells, co-generation, small gas turbines, and microturbines for distributed electricity generation.
EIA Natural Gas Hydrates	U.S. Energy Information Administration (EIA), chapter 3 "Future Supply Potential of Natural Gas Hydrates" in Energy Information Association, *Natural Gas 1998: Issues and Trends* (April 1998)	EIA describes the vast reserves of methane trapped in hydrated form in deep undersea and Arctic deposits, and discusses the technological prospects for recovery.
KPMG	KPMG, *Solar Energy: From Perennial Promise to Competitive Alternative* (August 1999)	A Dutch firm, KPMG, with the sponsorship of Greenpeace, proposes construction of large-scale (500 MW) photovoltaic power plants as a way of decreasing the cost of solar electricity.

Table 1—Continued

Model/Scenario	Source	Notes
EERE Program Plans	National Renewable Energy Laboratory (NREL), *Photovoltaics: Energy for the New Millenium* (January 2000)	NREL published several reports detailing the current state of federal renewables research. In the report on photovoltaics, NREL projects growth rates of photovoltaic systems and reductions in system costs, including an industry-developed roadmap with photovoltaics providing 10% of electricity by 2030. The Federal Wind Energy Program envisions prices of wind energy to fall to 2-4 cents by 2002. Further research and development could lower this price to 1-3 cents by 2015.

NOTE: These models and scenarios are widely used for energy policy planning purposes. Each model or scenario incorporates different assumptions about variables like fuel mix, political climate and economic change. The following table summarizes the sources and general characterization of the models and scenarios examined.

Three categories of parameters were used: *sociopolitical parameters*, *economic parameters*, and *energy parameters*. Note that 1996 constant dollars were used for economic measures, except as otherwise noted.

Sociopolitical Parameters

The types and quantities of energy needed and used depend both on the structure and behavior of society and on the existing energy supply and end-use infrastructure. Disruption occurs if required quantities of fuel are unavailable when needed, end-use patterns change rapidly or in unanticipated ways, or infrastructure breaks down. Such disruption can have economic impact (e.g., lost production), can cause personal inconvenience (e.g., power outages, long lines at the gas station), and can affect political decisions (e.g., trading partners, military alliances). An estimate of the possible level of disruption is an important sociopolitical metric that is difficult to quantify. Hence, a qualitative estimate was used to capture the potential for disruption as high, medium, or low.

The following are the sociopolitical parameters used as scenario descriptors in this study.

SP_1: Potential for disruption (high, medium, or low)

SP_2: Energy contribution to the Consumer Price Index (percent)

SP_3: Cost of health and environmental impacts and regulatory compliance ($/MBTU)

Most scenarios do not provide information on parameters SP_2 and SP_3. Nonetheless, these parameters were viewed as useful and informative. Many other parameters were considered, but the ones chosen were viewed as concise and relatively simple indicators of important sociopolitical effects of energy use in the economy.

Economic Parameters

The following parameters are used to describe the economic aspects of the scenarios.

EC_1: Gross domestic product (GDP) growth (percent per year)

EC_2: Inflation rate (percent per year)

EC_3: Energy price inflation/overall price inflation (ratio)

EC_4: Fuel taxes, energy subsidies, and R&D expenditures ($/MBTU)

Most scenarios provide information on GDP growth, but few provide information on inflation. Many scenarios do not provide full information on taxes, subsidies, and R&D expenditures. In some cases, policy surrogates provided a basis for estimating these parameters. Note that 1996 constant dollars were used for economic measures, as a matter of convenience in terms of the year selected, and to permit comparison of scenarios using a common metric.

Energy Parameters

Energy parameters used in this study characterize energy supply and demand, as well as the fuel mix and end use system.

EN_1: Total energy consumption (Quadrillion BTUs per year)

EN_2: Decarbonization[8] (dimensionless, with unity corresponding to exclusive coal use, and infinity corresponding to exclusive use of nonfossil fuels)

EN_3: Energy productivity of the economy ($ GDP/MBTU)[9]

"Full" energy scenarios provide data on all of these energy parameters. Scenarios that provide data on some, but not all, of the parameters, or provide incomplete data, are called "partial" scenarios. In addition to full and partial scenarios, some technology studies were analyzed because they provide useful input data for energy scenarios.

Analysis of Meta-Scenarios

The individual scenarios in Table 1 may be viewed as variations on four meta-scenarios. A fifth meta-scenario (Hard Times), characterized by slow economic growth and relatively low energy consumption, was added to provide a broader range for robust policy planning. This was intended to represent surprises and catastrophes that are not built into models, but could lead to very slow economic growth and stagnant energy consumption, as happened for example in the 1970s.

The five meta-scenarios are summarized in Table 2 and outlined below. They are arrayed according to a rough characterization of their economic growth rate and

[8]This is measured by the weighted sum of energy consumption per fuel type, normalized to total energy consumption, where the weights reflect the carbon emissions of each fuel per MBTU of energy consumed.

[9]This ratio is referred to as *energy productivity* to emphasize the fact that it includes more than the simple efficiency of electrical devices. It also includes the effects of sophisticated production and use choices that are increasingly available through the use of information technology, such as avoiding the production of excess inventory and using automated timers to control heating and air conditioning in buildings.

Table 2

Summary of Five Meta-Scenarios

Parameters	Hard Times	Business-as-Usual	Technological Improvement	High-Tech Future	New Society
Defined as:	Low growth/moderate environmental impact	Moderate growth/high environmental impact	Moderate growth/low environmental impact	Moderate growth/moderate environmental impact	Low growth/low environmental impact
Potential for Disruption (SP1)	High	Medium, because of the high level of oil imports and high reliance on natural gas, both of which pose potential problems of price increases and supply security or availability.	Medium, because of the need to substantially increase either energy productivity or decarbonization or combination of the two.	Medium, because of the high level of oil imports and high reliance on natural gas, both of which pose potential problems of price increases and supply security or availability.	High, because of the high level of policy intervention required to reach this future.
Energy Contribution to the Consumer Price Index (SP2)	Probably increased because of scarce energy and low economic growth.	Not addressed.	Objective would be to keep it the same as today	Objective would be to keep it the same as today	Low

Table 2—Continued

Parameters	Hard Times	Business-as-Usual	Technological Improvement	High-Tech Future	New Society
Cost of Health and Environmental Impacts and Regulatory Compliance (SP3)	Reduced because of lower energy consumption and lower economic growth, unless the pathway to this scenario was an environmental catastrophe, in which case these costs would be very large.	EPA has recently estimated the cost of regulation at $150-200 billion/year.	Should be reduced because of lower energy consumption and use of cleaner technology, as evidenced by increased decarbonization and energy productivity.	Higher energy consumption balanced by use of cleaner technology, as evidenced by increased decarbonization and energy productivity, could leave this the same as today.	Should be greatly reduced because of lower energy consumption and use of cleaner technology, as evidenced by increase decarbonization and energy productivity.
GDP Growth (EC1)	Close to zero	1.7-2.6%/year, with the EIA base case at 2.2.%/year.	2.2%/year (EIA Base Case)	3.2%/year (.5 higher than EIA high GDP variant).	1.7%/year
Inflation Rate (EC2)	A few percent or less	2.7%/year.	2.7%/year (EIA Base Case)	Not Addressed	Not Addressed
Energy Price Inflation /Overall Price Inflation (EC3)	Could be increased because of energy shortages.	Oil prices assumed to be in the range of $15-28/barrel in 2020.	Could increase due to higher fuel taxes.	Not Addressed	Not Addressed
Fuel Taxes, Energy Subsidies, and R&D Expenditures (EC4)	No major changes in taxes, but subsidies and R&D expenditures are reduced	Order of magnitude estimate is tens of billions of dollars per year, based upon available data and studies.	May require increased fuel taxes; will definitely require increased R&D expenditures.	May require increased energy subsidies; will probably require increased R&D expenditures.	Will require increased fuel taxes, removal of energy subsidies, and increased R&D expenditures.

Table 2—Continued

Parameters	Hard Times	Business-as-Usual	Technological Improvement	High-Tech Future	New Society
Total Energy Consumption (EN1)	100 quads	112-129 quads	97-110 quads	120-127 quads	65-83 quads
Decarbonization (EN2)	1.6	1.5-1.6 (1.61 in 1997)	1.5-2.0	1.6-1.9	1.8-2.6
Energy Productivity of the Economy (EN3)	$90/MBTU	$104-118/MBTU	$105-133/MBTU	$112-144/MBTU	$150-192/MBTU

their impact on the environment. Parameters are expressed in terms of increases and decreases with respect to the Business-as-Usual scenario.

Hard Times

(Lower Growth—Moderate Environmental Impact)

Under this meta-scenario, an economic downturn, supply constraint, or environmental catastrophe, or a combination of these events, leads to low- to zero- energy growth and no new infusion of new technologies. This implies very slow productivity growth and reliance on the same fuel mix.

There are many possible events that could trigger this sort of slowdown in energy use without much improvement in energy productivity such as: Middle East war, environmental catastrophe tied either by perception or reality to increased greenhouse gas emissions, and worldwide economic recession. None of the full energy scenarios fall into this category, because they do not consider these kinds of surprises or discontinuities. The Hard Times meta-scenario is similar to America's energy history between 1973 and 1984.

Signposts. The key indicator of Hard Times would be zero energy growth, little productivity growth, and a stagnant economy.

Shaping Strategies. This is an undesirable future, but inadvertent shaping strategies might include: heavily increased regulatory constraints on energy development; removal of incentives for increased energy productivity; and flawed policies leading to economic recession.

Hedging Strategies. In response to an uncertain future, hedging strategies could include increased R&D to improve energy productivity and renewable energy technologies; incentives for energy productivity and decarbonization; incentives for oil and gas exploration; and relicensing of nuclear power plants.

Business-as-Usual

(Moderate Growth—Higher Environmental Impact)

This meta-scenario extrapolates current trends in energy growth with continued improvement in energy productivity. However, the fuel mix actually becomes slightly more carbonized because use of nuclear power is decreasing and use of fossil fuels is increasing. This meta-scenario is clustered around the EIA *Annual Energy Outlook 2000* results.

There are many obstacles to reaching the Business-as-Usual future. This meta-scenario assumes that the U.S. will continue, simultaneously, to increase oil imports, increase use of natural gas (e.g., for essentially all new electric capacity additions), while using more coal, decommissioning nuclear plants on schedule, and making little progress on adoption of renewable energy alternatives. Price and security of oil supply, price and adequacy of gas supply, and acceptability of higher levels of carbon emissions are all uncertainties that could derail this extrapolation, especially with respect to the economic and sociopolitical parameter. For example, this meta-scenario could be invalidated through disruption, decreasing GDP growth, increasing the energy contribution to the consumer price index, or increasing costs of health and environmental impacts and regulatory compliance.

From 1973 to 1984, a combination of supply constraints and economic downturn kept energy growth constrained. Moreover, since 1985, the U.S. has been increasing energy productivity. This trend will likely continue as long as the U.S. continues to employ new technology at the current replacement rate.

Signposts. The signposts for this meta-scenario would likely include continued growth in energy demand with productivity increasing at its current rate, and decarbonization remaining constant or decreasing by 2010.

Shaping Strategies. Strategies could include incentives for increased oil and gas exploration, and support for emissions trading strategies.

Hedging Strategies:. Similar to Hard Times and the other meta-scenarios, these strategies would likely include increased R&D to improve energy productivity and renewable energy technologies; incentives for energy productivity and decarbonization; and relicensing of nuclear power plants.

Technological Improvement

(Moderate Growth—Lower Environmental Impact)

Under this meta-scenario, improvements would be expected in productivity and/or decarbonization as a result of improved technology. Moderate economic growth would occur with much smaller growth in energy consumption. The EIA Kyoto and both 5-lab Study full scenarios fall into this category.

Technological Improvement has several different possible pathways. Because the energy growth is modest, paths include going through a period of *Hard Times*, which is similar to the historical path to 2000. It also requires changes in productivity and decarbonization that are not too different from historical trends.

One possible business-as-usual pathway from the present might pass through a period in which there is little or no growth in energy consumption, but substantial increase in energy productivity (between 2000 and 2010). This might happen, for example, because of supply constraints or an environmental problem short of what it would take to put the U.S. into the *Hard Times* scenario. Within a business-as-usual set of assumptions, productivity would continue to improve at about the current rate, or perhaps somewhat faster to maintain some level of economic growth. Under this meta-scenario, recovery of fuel supplies or solution of the environmental problem—for example, new technology—might enable continued growth in both energy consumption and energy productivity (between 2010 and 2020). This is what has happened in the U.S. since 1985. The sociopolitical and economic parameters of such a scenario would depend strongly on the details of the pathway.

Signposts: Signposts would include near-term fuel shortages or greatly underestimated environmental impacts. A "glitch" in energy growth lasting more than a year or two, together with increased implementation of new energy technologies (e.g., hybrid vehicles, microturbines) would also be signposts of this meta-scenario.

Shaping Strategies: Intentional strategies could include incentives for energy productivity and decarbonization. Inadvertent shaping strategies might include rapidly escalating fuel prices or infrastructure failures.

Hedging Strategies: Hedging strategies could include incentives for oil and gas exploration, relicensing of nuclear power plants, increased R&D of energy productivity and renewable energy technologies, and incentives for energy productivity and decarbonization.

High Tech Future

(Higher Growth—Moderate Environmental Impact)

In the High Tech Future meta-scenario, economic and energy growth are similar to Business-as-Usual, but with technological advances that provide for productivity or decarbonization improvements like that are more like Technological Improvement. The Royal Dutch Shell Sustained Growth scenario and one of the World Energy Council High Growth scenarios fall into this category.

The High Tech Future has a combination of challenges. It requires energy growth like Business-as-Usual along with improvements in the fuel mix and energy productivity, approaching those described in the New Society scenario below.

Thus, to reach the High Tech Future, many of the obstacles to energy supply described under Business-as-Usual would need to be overcome, while also making major changes in energy use. However, higher economic growth could spawn large and rapid technical change and equipment turnover.

Signposts. Signposts could include continued economic prosperity leading to higher economic growth, increasing rate of adoption of new technology, such as hybrid vehicles, and an abundance of cheap oil and gas.

Shaping Strategies. Intentional strategies could include incentives for oil and gas exploration; R&D incentives and subsidies for energy productivity and renewable energy technologies; and relicensing of nuclear power plants.

Hedging Strategies. Hedging strategies would include increased R&D to improve energy productivity and renewable energy technologies, and incentives for energy productivity and decarbonization.

New Society

(Lower Growth—Benign Environmental Impact)

In the New Society meta-scenario, environmentally conscious and energy efficient choices in technology and lifestyle lead to much higher productivity, much higher decarbonization, and decreased energy consumption. The ACEEE and WEC Low Growth full energy scenarios fall in this category.

To reach the New Society, the U.S. would have to do something it has never done before: reduce total energy consumption (from approximately 100 quads to 65–83). The scales of productivity and decarbonization improvements are daunting compared to those of the past, even with the downturn in economic growth embodied in this scenario:.

- Energy productivity has increased from $54/MBTU to $92/MBTU in the past 40 years, an average of 1.8%/year. New Society requires an increase from $92 to $150–192/MBTU in the next 20 years, an average of 3.1–5.6% per year.
- Decarbonization has increased from 1.49 to 1.62 over the past 40 years, an average of 0.2% per year. New Society requires an increase from 1.62 to 1.8–2.6 in the next 20 years, an average of 0.5–2.3% per year.

To achieve reduced energy consumption *and* a less carbon-intensive fuel mix at the same time, the U.S. would need to change its energy use in a revolutionary way. Technologies such as fuel cells, photovoltaics, electric and hybrid vehicles are presently much more expensive than current alternatives. Very large resources would be needed to pay these costs, for example, in the form of increased R&D investment, increased subsidies or incentives for energy productivity and renewable energy technologies, and/or increased fuel taxes. If successful, health and environmental benefits would accrue from decreased energy use and use of cleaner energy technologies. A more complete analysis of this meta-scenario should evaluate the rate of turnover of energy conversion and utilization equipment, infrastructure improvements and modifications (for example, in electricity storage), and the possible time to implement any necessary lifestyle changes (such as land use, public transportation, and work patterns). Such analysis would indicate the degree of difficulty in actually getting to this future in 20 years.

Signposts: The key signposts would likely be revolutionary increases in energy productivity and decarbonization; accelerated use of renewable energy technologies; and lifestyle changes involving greater use of mass transit, less driving, and load-leveling electricity use.

Shaping Strategies. Major emphasis and resources would need to be devoted to energy productivity improvement and accelerated adoption of renewable energy technologies, including incentives, subsidies, and public education.

Hedging Strategies. Hedging strategies would include incentives for oil and gas exploration, R&D on clean coal technologies, and relicensing of nuclear power plants.

Policy Implications

This exercise is suggestive of the kinds of policy implications that could be drawn from a more extensive strategic planning exercise that sought pathways to more desirable energy futures. For example, it is clear and unsurprising that research and development serves as a hedging strategy under all scenarios. Because technological improvement and market penetration are all subject to uncertainty, a diverse portfolio of research and development efforts is advisable, as recommended by PCAST.

Another example is the ramifications of some policy choices on natural gas use. If decarbonization policies are pursued, many of the scenarios examined suggest natural gas as a transition fuel, whether the gas is used by fuel cells or advanced combined-cycle turbines. For example, Jesse Ausubel's long range study of historical energy trends and their implications for the future suggests that natural gas could be on the brink of a take-off, mirroring earlier experiences with coal and oil. Decisionmakers will need to ensure that policy choices fully reflect and respond to such a change from the current situation.

It is also clear that much more needs to be understood about how the Internet and information technology are changing energy use, and possibly resulting in some of the most fundamental change ever seen. The E-Vision 2000 Policy Forum panel discussion on information technologies and energy use suggest the dimensions of the R&D effort required to grasp the full implications of this IT/energy connection.

Finally, new technologies represent another wildcard in any energy visioning process. For example, the partial scenario provided by Lovins and Williams, and other technology studies of distributed generation (ADL) and fuel cells (CARB), suggest that some technologies may have a major transformational effect on the entire energy system. For example, it may be argued that because fuel cells have such a range of potential applications (remote, reliable, uninterrupted stationary power today, distributed generation tomorrow, then automotive use), they may ultimately enable transition to a hydrogen economy.

Conclusions

In summary, the E-Vision 2000 scenario analysis exercise highlighted the need to consider disruptions and shocks to the status quo, particularly with regard to fossil energy sources; the value of better understanding realistic rates of adoption of new technologies; and the need to analyze the various options available with regard to nuclear power. None of the planning scenarios examined in this analysis explored the effect of "surprises," for example, oil price "shocks" such as occurred in 1973 and 1979. Whatever the future holds in store, it is likely to include a lot less continuity than depicted in these scenarios or even the Hard Times meta-scenario. Indeed, the historical record of U.S. energy consumption shows periods of rapid growth, crisis and adjustment, and then continued growth, albeit at a slower rate, and accompanied by increased energy efficiency.

Hedging Against Possible Supply Disruptions

A significant group of scenarios assumes that the U.S. will not significantly change its fuel mix by 2020, implying increased use of coal. Unanticipated events—such as Middle East war, global climate catastrophe (real or perceived), failure to obtain approvals for new coal-fired power plants, failure to add to gas reserves at anticipated rates, liquefied natural gas or oil tanker accidents—could lead to future energy supply constrictions. Aggressive policy actions to increase the fraction of cleaner, domestic energy in the fuel mix and increase energy efficiency are the best hedge possible against such futures.

Availability and Security of Fossil Fuel Sources

Because most scenarios, especially EIA's, show increased use of oil and natural gas, the source and security of oil and gas supply, including imports, is a key policy issue. Increased oil imports are assumed to come primarily from the Persian Gulf. Increases in domestic supply of natural gas are assumed to come from additions to proven reserves. Alternative oil or liquid fuel supply options and necessary price and policy incentives for natural gas production are critical issues that require analysis.

Feasibility of Rapid Technological Change

Many of these scenarios assume rapid adoption of improved technology, sometimes coupled with changes in patterns of energy consumption. These futures are unlikely to come about without aggressive policy action, such as the type of broad energy R&D and deployment program envisioned by PCAST, or a greenhouse gas emissions "cap and trade" program. It is by no means clear that such a future is obtainable through pursuit of existing policies. In fact, such policies, along with new fuel discoveries like methane hydrates, and a sustained economic boom, could well be driving forces toward increased growth of energy use.

Need for Analysis of Nuclear Options

Most scenarios wrote off the nuclear option by assuming that nuclear power plants will be decommissioned on schedule and that no nuclear power plants will be built. While this is consistent with current trends in the U.S. and Europe, it is too narrow an assumption to inform policy. As demonstrated in the EIA Kyoto Protocol scenario variants, extending the lifetime of existing plants can be an essential and cost-effective component of a carbon reduction strategy. Especially in light of the current level of international concern about greenhouse gases, nuclear power, as a carbon-free source of electricity, needs to be analyzed and considered within an objective framework that compares the costs, risks, and impacts of alternative energy sources.

Policy Questions

From a policy perspective, the difficult step in analysis is to take these scenarios and begin to map out pathways to implementing strategies that would most likely lead to more desirable futures, especially those that lead to a cleaner fuel mix and more efficient energy consumption. Research questions include:

- What is the fuel mix along the way?
- What sectors are affected and how?
- What policy choices must be made?
- Is there, in fact, any way to get there from current conditions?

Answering these questions is a predicate for taking the next step to devising robust and effective strategies for deploying R&D resources.

Expert Opinion on Key Energy Issues in 2020

James Dewar
RAND

John Friel
RAND

Twenty-seven high-level national energy thinkers participated in a process to identify the most important issues that will affect energy trends in the next 20 years. The process used is called the "modified Delphi method," developed at RAND from studies on decisionmaking that began more than 50 years ago. The Delphi method is designed to take advantage of the talent, experience, and knowledge of a group of experts in a structure that allows for an exchange of divergent views. In this application, EERE was seeking a formal means of capturing expert opinion of energy needs in 2020, which would inform R&D portfolio choices over the next five to ten years.

The Delphi technique and specific results of this process were presented at the E-Vision 2000 conference. This section summarizes the methodology and results of the Delphi process, and identifies ways in which the process could be improved for future applications.[10]

Approach and Methodology

The Delphi method is used in strategic planning to project future technical, market, and other developments, uncover fundamental differences of opinion, and identify non-conventional ideas and concepts (Technology Futures, Inc.). Participants first make initial projections of future events. After their initial projections are correlated and shared with the group, participants are then asked to explain (anonymously) their differences in a series of follow-up rounds.

The Delphi method (Dalkey and Helmer) was developed as an alternative to the traditional method of obtaining group opinion through face-to-face discussions.

[10]See http://www.rand.org/scitech/stpi/Evision/Transcripts/Day1-am.pdf for the relevant section of the conference transcripts. See http://www.rand.org/scitech/stpi/Evision/Supplement/delphi.pdf for the slides from James Dewar's presentation.

Experimental studies have demonstrated several serious difficulties with such discussions, including:

- Influence of a dominant individual. The group is highly influenced by a person who talked the most or has most authority;
- Group interest dynamics. Studies found that "communication" in face-to-face discussions often had to do with individual and group interests rather than problem solving;
- Group pressure for conformity. Studies demonstrated the distortions of individual judgment that can occur from group pressure.

In its original formulation, the Delphi method had three basic features (Dewar and Friel, 1996):

- Anonymous response. Opinions of group members are obtained by questionnaire;
- Iteration and controlled feedback. Interaction is effected by a systematic exercise conducted in several iterations with carefully controlled feedback between rounds; and
- Statistical group responses. The group opinion is defined as an appropriate aggregate of individual opinions on the final round.

Application of the Delphi Method to Energy Needs

For this EERE study, a more qualitative variation of the Delphi method was used. RAND asked experts for opinions rather than specific numeric projections. This "modified Delphi technique" still retained the anonymity of participants as well as iteration of responses and specific feedback. However, the statistical group response was dropped because of the qualitative nature of the question. The exercise was conducted entirely over the Internet.

RAND developed a scenario about the future and the participants were asked to respond with a set of "yes/no" questions. The scenario was as follows:

A time traveler from 20 years in the future will visit you early in the new millennium. The time traveler knows everything about the situation surrounding energy needs in the year 2020. What do you want to know about the future?

In the first round of this iterative process, each participant could submit ten questions about the future energy scene; and those responses were fed back to the entire group. Participants could revise their initial questions and comment on other responses as part of Round 2. In Round 3, participants were limited to six questions rather than ten in order to have them focus on top priority issues.

The participants in the Delphi exercise represented three broad categories: private sector/industry trade associations, government, and academia/nonprofit organizations. Table 3 lists the various organizations of individuals who participated in the process.

Table 3

Organizations of Participants in the Delphi Process

Organization
Private Sector and Industry Trade Associations
American Electric Power
CONSAD Research Corporation
Ford Motor Company
Gas Research Institute
Lehr Associates
McDonough & Associates
Robert Charles Lehr & Co.
SAIC
Weyerhaeuser
Federal Government
National Energy Technology Laboratory
National Renewable Energy Laboratory
Oak Ridge National Laboratory
Former Presidential Science Advisor
State Energy Advisory Board (DOE)
Academia and Nonprofit Organizations
Carnegie Mellon University
Colorado School of Mines
James Madison University
Purdue University
University of Houston/Energy Institute
American Council for an Energy-Efficient Economy Energy Foundation
Global Environment & Technology Foundation/Center for Energy and Climate Solutions
International Research Center for Energy an Economic Development, Sigma Xi

Summary of Results

The RAND project team examined the responses from each round of the Delphi process and organized the results in terms of consensus development within different categories of participants and issue areas.

In Round 1, the 27 participants posed 262 questions (out of 270 maximum); some participants did not submit their full complement of 10 questions. In Round 2, the participants had the opportunity to select from among the original questions or submit new questions, with the limit of 10 questions per participant. By the second round, there were 185 questions out of a potential maximum of 270. In Round 3, participants were limited to six questions: 99 different questions were submitted out of a possible 150 (only 25 participants responded in Round 3). These summary results show some convergence, but little consensus on the actual set of questions.

Questions Clustered by Issue Area

When questions are clustered by general topic, there was much greater consensus among participants than observed in the aggregate. The results indicate the following issue areas attracted the most interest:

- Global warming/greenhouse gas emissions (70 percent of the participants)
- Hybrid/zero-emission vehicle market penetration (59 percent of the participants)
- Natural gas usability/use (56 percent of the participants)
- Increased nuclear use (33 percent of the participants
- Oil prices (26 percent of the participants)
- Fuel cell viability (19 percent of the participants)

E-Vision Conference Issues

Prior to the E-Vision 2000 conference, EERE identified its priority future energy issues. Three of the topics—information technology, worker productivity, and systems issues—became the focus of the conference itself. We examined the views of the Delphi participants on the remaining EERE questions. The results show there was considerable interest in environmental issues (89% of participants) and transportation issues (59% of participants), but moderate to low priority given to the other top EERE issues. The topics and number of participants follow.

Conclusion

The Delphi participants clearly viewed global warming and carbon dioxide emission controls as the most important issues affecting future energy trends. Among technologies, natural gas electricity generation, fuel cells and nuclear power drew the greatest interest. In addition, there was clear support for EERE's concerns with future roles of transportation and environmental issues.

This particular application of the Delphi method failed to produce clear convergence around a limited set of issues. This was due in part to the qualitative nature of the responses. The question itself failed to stimulate "out of the box" thinking, but rather produced concerns consistent with the current debate. In the future, non–energy experts from a wide range of fields will be engaged in envisioning different energy futures and associated R&D investments implicated by those possible scenarios. The results of this Delphi exercise will be used to focus the planning of future research efforts as well as subsequent E-Vision conferences.

References

Dalkey, N. and O. Helmer (1963), "An Experimental Application of the Delphi Method to the Use of Experts," *Management Science*, 9, 458–467

Dewar, James A. and John A. Friel (1996), "Delphi Method" in S.I. Gass and C.M. Harris, eds., *Encyclopedia of Operations Research and Management Science*, Kluwer.

Technology Futures, Inc. (TFI), "Forecasting Techniques," Internet: http://www.tifi.com (last accessed March 2001).

Abstracts of Submitted Papers

As part of the preparation for the E-Vision 2000 Conference, RAND commissioned papers on behalf of EERE on a wide range of topics. These were distributed in advance of the conference and served as points of departure or reference for the panel discussions and open forum. These papers represent the views of the authors and not those of RAND or the Department of Energy. The abstracts below have been arranged by panel topic.[11]

[11]See http://www.rand.org/scitech/stpi/Evision/Supplement/ for the full texts in PDF form of all the submitted papers. Subsequent footnotes provide the precise link for each paper.

The Influence of Information Technologies on Energy Use

Information Technology Impacts on the U.S. Energy Demand Profile

Brad Allenby
Vice President, Environment, Health and Safety, AT&T

Darian Unger
Researcher, Massachusetts Institute of Technology

The growth of information technology (IT) and the Internet may dramatically change society and the energy sector. Effective energy policies require sound research and development (R&D) to investigate these changes. This paper explains how viewing IT and energy in the context of a social system allows us to recognize that the indirect and unexpected effects of IT may dwarf the direct impacts on energy consumption. It also makes recommendations for the difficult task of divining the future of energy demand.[12]

Competitive Electricity Markets and Innovative Technologies: Hourly Pricing Can Pave the Way for the Introduction of Technology and Innovation

William L. Reed
Vice President and Chief Regulatory Officer, Sempra Energy Corporation

For most electricity consumers, the transition to a competitive market has been unremarkable—if noticed at all—because the transition has come almost entirely at the wholesale level, where few consumers are involved. Not much has changed in the retail market for electricity or in the consumer's understanding of electricity services to make for effective competition. As an early proponent of a competitive retail electricity market, Sempra Energy advocates the imposition of hourly retail electricity prices. This is an essential step to create competition in the retail end of the electricity market and, ultimately, more effective competition throughout the electricity sector.[13]

---

[12]Full text: http://www.rand.org/scitech/stpi/Evision/Supplement/allenby.pdf

[13]Full text: http://www.rand.org/scitech/stpi/Evision/Supplement/reed.pdf

Raising the Speed Limit: U.S. Economic Growth in the Information Age

Dale W. Jorgenson
Director, Program on Technology and Economic Policy, John F. Kennedy School of Government, Harvard University

Kevin J. Stiroh
Senior Economist, Federal Reserve Bank of New York

This paper examines the underpinnings of the successful performance of the U.S. economy in the late 1990s. Relative to the early 1990s, output growth has accelerated by nearly two percentage points. This growth is attributed to rapid capital accumulation, a surge in hours worked, and faster growth of total factor productivity. The acceleration of productivity growth, driven by information technology, is the most remarkable feature of the U.S. growth resurgence. This paper considers the implications of these developments for the future growth of the U.S. economy.[14]

The Internet and the New Energy Economy

Joseph Romm
Executive Director, Center for Energy and Climate Solutions, Global Environment and Technology Foundation

From 1996 through 1999, the U.S. experienced an unprecedented 3.2% annual reduction in energy intensity, four times the rate of the previous 10 years and more than 3 times higher than the rate projected by traditional energy forecasters. There is increasing data and analysis to support the view that there is a connection between the recent reductions in energy intensity and the astonishing growth in Information Technology (IT) and the Internet Economy. Growth in the Internet Economy can cut energy intensity in two ways. First, the IT sector is less energy-intensive than traditional manufacturing, so growth in this sector engenders less incremental energy consumption. Second, the Internet Economy appears to be increasing efficiency and productivity in every sector of the economy, which is the primary focus of this paper. The impact of the Internet economy on manufacturing, buildings, and transportation are all explored. The paper also considers the implications for growth in energy consumption and greenhouse gas emissions during the next ten years. Finally, the paper disputes an argument put forward by two analysts, Mark Mills and Peter Huber, that the Internet is using a large and rapidly growing share of the nation's electricity,

[14]Full text: http://www.rand.org/scitech/stpi/Evision/Supplement/jorgenson.pdf

which in turn is supposedly driving an acceleration of overall U.S. electricity demand.[15]

Power for a Digital Society

Kurt Yeager
President, Electric Power Research Institute

Karl Stahlkopf
Vice President, Power Delivery, Electric Power Research Institute

Once or twice in a century, "enabling sectors" of the economy emerge which are so profound in their impact that they transform all other sectors of the economy. In the 21st century, information networks relying upon integrated circuits (microprocessors) powered by electricity and linked by high-speed, broadband communication are envisioned to have a transformative role in creating the Digital Society. This technology revolution has progressed through three stages, each one more significant in its impact on society. Increasingly, stand-alone microprocessors are being linked to networks, supplying critical information on equipment operations and facilitating even more profound changes in daily life.[16]

Worker Productivity, Building Design, and Energy Use

Health and Productivity Gains from Better Indoor Environments and Their Implications for the U.S. Department of Energy

William J. Fisk
Staff Scientist, Lawrence Berkeley National Laboratory

A substantial portion of the U.S. population suffers frequently from communicable respiratory illnesses, allergy and asthma symptoms, and sick building syndrome symptoms. There is now increasingly strong evidence that changes in building design, operation, and maintenance can significantly reduce these illnesses. Decreasing the prevalence or severity of these health effects would lead to lower health care costs, reduced sick leave, and shorter periods of illness-impaired work performance, resulting in annual economic benefits for the U.S. in the tens of billions of dollars. Increasing the awareness of these potential health and economic gains, combined with other factors, could help bring about a shift in the way we design, construct, operate, and occupy buildings.

---

[15]Full text: http://www.rand.org/scitech/stpi/Evision/Supplement/romm.pdf

[16]Full text: http://www.rand.org/scitech/stpi/Evision/Supplement/yeager.pdf

Additionally, DOE's energy-efficiency interests would be best served by a program that prepares for the potential shift, specifically by identifying and promoting the most energy-efficient methods of improving the indoor environment. The associated research and technology transfer topics of particular relevance to DOE are identified and discussed.[17]

The Urban Structure and Personal Travel: An Analysis of Portland, Oregon, Data and Some National and International Data

T. Keith Lawton
Director of Travel Forecasting, Department of Transportation, City of Portland, Oregon

In Portland, Oregon, there has been a desire to understand the mechanisms leading to travel demand, and the part played by urban design and the configuration of land uses in the urban region. Portland data show that residents in denser, inner-city, mixed use areas consume significantly fewer miles of car travel, substituting slow modes, such as walking, bicycling and transit. Primarily lower-order (slow) streets serve these inner city dwellers. These have some congestion, so that car mobility, expressed as speed, is sharply reduced. Yet the data suggest that these people spend much the same time traveling, and have as many out of home activities as their suburban cousins. Most of this plan evaluation in Portland is carried out using transportation models developed from a household activity and travel survey carried out in the Portland metropolitan area in 1994 and 1995. The implications of data from this household activity and travel survey are discussed in the paper.[18]

Enhancing Productivity While Reducing Energy Use in Buildings

David P. Wyon
Research Fellow and Adjunct Professor, Johnson Controls, Inc. and Technical University of Denmark

The problem addressed in this paper is the conflict between the undoubted need to reduce energy use in buildings and the reasonable economic requirement that energy conservation initiatives should not negatively affect indoor environmental quality in offices and schools and diminish productivity. Indoor air quality (IAQ) and air temperature (T) have powerful effects on the efficiency with which work can be performed in schools and offices. Huge amounts of energy are used to keep these parameters constant at levels which represent a

---

[17]Full text: http://www.rand.org/scitech/stpi/Evision/Supplement/fisk.pdf

[18]Full text: http://www.rand.org/scitech/stpi/Evision/Supplement/lawton.pdf

compromise between group average requirements for subjective comfort and energy conservation. The national economic interest would be served by the establishment of a virtual institute to actively apply and orchestrate research to establish the viability of proposed solutions to the conflict between energy conservation and productivity.[19]

Systems Approach to Energy Use

Promoting Renewable Energy and Demand-Side Management Through Emissions Trading Program Design

Jack Cogen
President, Natsource

Matthew Varilek
Analyst, Advisory Services Unit, Natsource

This paper discusses the integration of incentives for renewables generation and demand-side management within the context of emissions trading markets such as the U.S. Sulfur Dioxide Allowance Trading Program and the emerging market for greenhouse gas emissions reductions. As the popularity of emissions trading continues its rapid growth, so also can opportunities for promotion of renewable energy and demand-side management, which undoubtedly contribute to achievement of emissions trading programs' objectives. But this promise will only be realized if deficiencies in the design of existing trading programs can be rectified, if not within existing programs, then in emerging ones. In this paper, deficiencies in existing programs are examined, and effective solutions are discussed.[20]

On the Road Again—In the Car of the Future

Josephine Cooper
President and CEO, Alliance of Automobile Manufacturers

The Alliance of Automobile Manufacturers is a coalition of 13 car and light truck manufacturers, including BMW Group, DaimlerChrysler, Fiat, Ford Motor Company, General Motors, Isuzu, Mazda, Mitsubishi Motors, Nissan, Porsche, Toyota, Volkswagen, and Volvo. Alliance members represent more than 90 percent of U.S. vehicle sales. The automakers' vision for the future is premised on achieving technological breakthroughs for vehicles. These breakthroughs will

---

[19]Full text: http://www.rand.org/scitech/stpi/Evision/Supplement/wyon.pdf

[20]Full text: http://www.rand.org/scitech/stpi/Evision/Supplement/cogen.pdf

embrace additional environmental improvements, enhanced fuel efficiency and vehicle safety. New technologies must be acceptable to the market in all areas, including price, utility, and performance. The key to development of advanced technology vehicles lies in overcoming not only the hurdles and obstacles to their commercialization, but also acceptance by consumers. Creating the necessary infrastructure to support advanced technology vehicles is another critical factor in bringing vehicles to market.[21]

Energy R&D Policy in a Changing Global Environment

Charles K. Ebinger
Vice President, International Energy Services, Nexant

This slide presentation focused on the challenges of formulating an effective R&D policy in a changing market environment. Ebinger examined the demographic and institutional context for various energy futures, and in particular, considered the implications of changes spurred by the Internet and e-procurement.[22]

Taking the Time to ask the Right Questions or, CSTRR: A Case Study with Attitude

Rose McKinney-James
President, Government Affairs, Faiss Foley Merica,

During the past year and a half, the nation's attention has been directed toward the growing need to address existing energy policy. Recently, the national political debate has focused on the need to establish a comprehensive energy policy for the country. A substantial amount of attention has been given to the need to balance the interests of the array of stakeholders interested in influencing this policy. This paper focuses on the need to insure the integration of renewable energy resources as a foundation for national policy. It is based in part on the author's experience as a solar advocate in the state of Nevada. It offers one perspective on how energy other advocates might revisit past approaches and seize the ever-elusive "window of opportunity."[23]

---

[21]Full text: http://www.rand.org/scitech/stpi/Evision/Supplement/cooper.pdf
[22]Full text: http://www.rand.org/scitech/stpi/Evision/Supplement/ebinger.pdf
[23]Full text: http://www.rand.org/scitech/stpi/Evision/Supplement/mckinney.pdf

Distributed Generation Architecture and Control

Robert C. Sonderegger
Director of Modeling and Simulation, Silicon Energy Corporation

The traditional roles of utilities as energy suppliers, and consumers as energy users, are morphing rapidly. The confluence of deregulation with advances in telecommunications and information technology has opened new opportunities to decentralize generation. For example, power generation is no longer limited to centralized power plants. The technology is available to generate power at all levels, whether at the transmission, distribution, or at the end user levels. As the interconnections become more complex, and many actors produce and consume power simultaneously, traditional Supervisory Control and Data Acquisition (SCADA) is inadequate to handle, let alone optimize, all possible combinations of power production and consumption. This paper reviews some of the control issues arising from this brave new world of distributed power generation and some of the technologies that make its control possible.[24]

---

[24]Full text: http://www.rand.org/scitech/stpi/Evision/Supplement/sonderegger.pdf